T0270798

Practitioner's Guide to Data Science

This book aims to increase the visibility of data science in real-world, which differs from what you learn from a typical textbook. Many aspects of day-to-day data science work are almost absent from conventional statistics, machine learning, and data science curriculum. Yet these activities account for a considerable share of the time and effort for data professionals in the industry. Based on industry experience, this book outlines real-world scenarios and discusses pitfalls that data science practitioners should avoid. It also covers the big data cloud platform and the art of data science, such as soft skills. The authors use R as the primary tool and provide code for both R and Python.

This book is for readers who want to explore possible career paths and eventually become data scientists. This book comprehensively introduces various data science fields, soft and programming skills in data science projects, and potential career paths. Traditional data-related practitioners such as statisticians, business analysts, and data analysts will find this book helpful in expanding their skills for future data science careers. Undergraduate and graduate students from analytics-related areas will find this book beneficial to learn real-world data science applications. Non-mathematical readers will appreciate the reproducibility of the companion R and python codes.

Key Features:

- It covers both technical and soft skills.
- It has a chapter dedicated to the big data cloud environment. For industry applications, the practice of data science is often in such an environment.
- It is hands-on. We provide the data and **repeatable** R and Python code in notebooks. Readers can repeat the analysis in the book using the data and code provided. We also suggest that readers modify the notebook to perform analyses with their data and problems, if possible. The best way to learn data science is to do it!

CHAPMAN & HALL/CRC DATA SCIENCE SERIES

Reflecting the interdisciplinary nature of the field, this book series brings together researchers, practitioners, and instructors from statistics, computer science, machine learning, and analytics. The series will publish cutting-edge research, industry applications, and textbooks in data science.

The inclusion of concrete examples, applications, and methods is highly encouraged. The scope of the series includes titles in the areas of machine learning, pattern recognition, predictive analytics, business analytics, Big Data, visualization, programming, software, learning analytics, data wrangling, interactive graphics, and reproducible research.

Published Titles

Statistical Foundations of Data Science
Jianqing Fan, Runze Li, Cun-Hui Zhang, and Hui Zou

A Tour of Data Science: Learn R and Python in Parallel
Nailong Zhang

Explanatory Model Analysis
Explore, Explain, and, Examine Predictive Models
Przemyslaw Biecek, Tomasz Burzykowski

An Introduction to IoT Analytics
Harry G. Perros

Data Analytics
A Small Data Approach
Shuai Huang and Houtao Deng

Public Policy Analytics
Code and Context for Data Science in Government
Ken Steif

Supervised Machine Learning for Text Analysis in R
Emil Hvitfeldt and Julia Silge

Massive Graph Analytics
Edited by David Bader

Data Science
An Introduction
Tiffany-Anne Timbers, Trevor Campbell and Melissa Lee

Tree-Based Methods
A Practical Introduction with Applications in R
Brandon M. Greenwell

Urban Informatics
Using Big Data to Understand and Serve Communities
Daniel T. O'Brien

Introduction to Environmental Data Science
Jerry Douglas Davis

Hands-On Data Science for Librarians
Sarah Lin and Dorris Scott

Geographic Data Science with R
Visualizing and Analyzing Environmental Change
Michael C. Wimberly

Practitioner's Guide to Data Science
Hui Lin and Ming Li

For more information about this series, please visit: https://www.routledge.com/Chapman--HallCRC-Data-Science-Series/book-series/CHDSS

Practitioner's Guide to Data Science

Hui Lin
Ming Li

CRC Press
Taylor & Francis Group
Boca Raton London New York

CRC Press is an imprint of the
Taylor & Francis Group, an **informa** business

A CHAPMAN & HALL BOOK

Designed cover image: © Hui Lin and Ming Li

First edition published 2023
by CRC Press
6000 Broken Sound Parkway NW, Suite 300, Boca Raton, FL 33487-2742

and by CRC Press
4 Park Square, Milton Park, Abingdon, Oxon, OX14 4RN

CRC Press is an imprint of Taylor & Francis Group, LLC

© 2023 Hui Lin and Ming Li

Reasonable efforts have been made to publish reliable data and information, but the author and publisher cannot assume responsibility for the validity of all materials or the consequences of their use. The authors and publishers have attempted to trace the copyright holders of all material reproduced in this publication and apologize to copyright holders if permission to publish in this form has not been obtained. If any copyright material has not been acknowledged please write and let us know so we may rectify in any future reprint.

Except as permitted under U.S. Copyright Law, no part of this book may be reprinted, reproduced, transmitted, or utilized in any form by any electronic, mechanical, or other means, now known or hereafter invented, including photocopying, microfilming, and recording, or in any information storage or retrieval system, without written permission from the publishers.

For permission to photocopy or use material electronically from this work, access www.copyright.com or contact the Copyright Clearance Center, Inc. (CCC), 222 Rosewood Drive, Danvers, MA 01923, 978-750-8400. For works that are not available on CCC please contact mpkbookspermissions@tandf.co.uk

Trademark notice: Product or corporate names may be trademarks or registered trademarks and are used only for identification and explanation without intent to infringe.

Library of Congress Cataloging-in-Publication Data

Names: Lin, Hui (Quantitative researcher), author. | Li, Ming (Research
science manager), author.
Title: Practitioner's guide to data science / Hui Lin and Ming Li.
Description: First edition. | Boca Raton : CRC Press, 2023. | Includes
bibliographical references and index. | Identifiers: LCCN 2022052592 (print) |
LCCN 2022052593 (ebook) | ISBN
9780815354475 (hardback) | ISBN 9780815354390 (paperback) | ISBN
9781351132916 (ebook)
Subjects: LCSH: Big data. | Data mining. | Database management.
Classification: LCC QA76.9.B45 L56 2023 (print) | LCC QA76.9.B45 (ebook)
| DDC 005.7--dc23/eng/20230125
LC record available at https://lccn.loc.gov/2022052592
LC ebook record available at https://lccn.loc.gov/2022052593

ISBN: 978-0-815-35447-5 (hbk)
ISBN: 978-0-815-35439-0 (pbk)
ISBN: 978-1-351-13291-6 (ebk)

DOI: 10.1201/9781351132916

Typeset in Latin Modern font
by KnowledgeWorks Global Ltd.

Publisher's note: This book has been prepared from camera-ready copy provided by the authors.

Contents

List of Figures

Preface

In the early years of our data science career, we were bewildered by all the hype surrounding the field. There was – and still is – a lack of definition of many basic terminologies such as "big data," "artificial intelligence," and "data science." How big is big data? What is data science? What is the difference between the sexy title "Data Scientist" and the traditional "Data Analyst?" The term data science stirs so many associations such as machine learning (ML), deep learning (DL), data mining, and pattern recognition. All those struck us as confusing and vague as real-world data scientists!

However, we could always sense something tangible in data science applications, and it has been developing very fast. After applying data science for many years, we now have a much better idea about data science in general. This book is our endeavor to make data science a more concrete and legitimate field. In addition to the "hard" technical aspects, the book also covers soft skills and career development in data science.

Goal of the Book

This is a book on data science with a specific focus on industrial experience. Data Science is a cross-disciplinary subject involving hands-on experience and business problem-solving exposures. The majority of existing introduction books on data science are about modeling techniques and the implementation of models using R or Python. However, many of these books lack the context of the industrial environment. Moreover, a crucial part, the art of data science in practice, is often missing. This book intends to fill the gap.

Some key features of this book are as follows:

- It covers both technical and soft skills.
- It has a chapter dedicated to the big data cloud environment. In the industry, the practice of data science is often in such an environment.
- It is hands-on. We provide the data and **repeatable** R and Python code in notebooks. Readers can repeat the analysis in the book using the data and code provided. We also suggest that readers modify the notebook to perform their analyses with their data and problems whenever possible. The best way to learn data science is to do it!
- It focuses on the skills needed to solve real-world industrial problems rather than an academic book.

What This Book Covers

Numerous books on data science exist, yet few provide a comprehensive overview of both the technical and practical aspects. This book provides a comprehensive introduction to various data science fields, soft and programming skills needed for data science projects, and potential career paths. It is organized as follows:

- **Chapters 1–3** discuss various aspects of data science: difference tracks, career paths, project cycles, soft skills, and common pitfalls. Chapter 3 is an overview of the data sets used in the book.
- **Chapter 4** introduces typical big data cloud platforms and uses R library `sparklyr` as an interface to the big data analytics engine Spark.
- **Chapters 5–6** cover the essential skills to prepare the data for further analysis and modeling, i.e., data preprocessing and wrangling.
- **Chapter 7** illustrates the practical aspects of model tuning. It covers different types of model error, sources of model error, hyperparameter tuning, how to set up your data, and how to

make sure your model implementation is correct. In practice, applying machine learning is a highly iterative process. We discuss this before introducing the machine learning algorithm because it applies to nearly all models. You will use cross-validation or training/developing/testing split to tune the models presented in later chapters.

- **Chapters 8–12** introduce different types of models. There is a myriad of learning algorithms to learn the data patterns. This book doesn't cover all of them but presents the most common ones or the foundational methods.

Who This Book Is For

This book is for readers who want to explore potential data science career paths and eventually want to become a data scientist. Traditional data-related practitioners such as statisticians, business analysts, and data analysts will find this book helpful in expanding their skills for future data science careers. Undergraduate and graduate students from analytics-related areas will find this book beneficial to learn real-world data science applications. Non-mathematical readers will appreciate the reproducibility of the companion R and python codes.

How to Use This Book

What the Book Assumes

The first two chapters do not have any prerequisite, and the rest of the chapters do require R or Python programming experience and undergraduate level statistics. This book does NOT try to teach the readers to program in the basic sense. It assumes that readers have experience with R or Python. If you are new to programming languages, you may find the code obscure. We provide some references in the **Complementary Reading** section that can help you fill the gap.

For some chapters (5, 7–12), readers need to know elementary linear algebra (such as matrix manipulations) and understand basic statistical concepts (such as correlation and simple linear regression). While the book is biased against complex equations, a mathematical background is good for the deep dive under the hood mechanism for advanced topics behind applications.

How to Run R and Python Code

This book uses R in the main text and provides most of the Python codes on GitHub.

Use R code. You should be able to repeat the R code in your local R console or RStudio in all the chapters except for Chapter 4. The code in each chapter is self-sufficient, and you don't need to run the code in previous chapters first to run the code in the current chapter. For code within a chapter, you do need to run from the beginning. At the beginning of each chapter, there is a code block for installing and loading all required packages. We also provide the `.rmd` notebooks that include the code to make it easier for you to repeat the code. Refer to this page `http://bit.ly/3r7cV4s` for a table with the links to the notebooks.

To repeat the code on big data and cloud platform part in Chapter 4, you need to use Databricks, a cloud data platform. We use Databricks because:

- It provides a user-friendly web-based notebook environment that can create a Spark cluster on the fly to run R/Python/Scala/SQL scripts.
- It has a free community edition that is convenient for teaching purpose.

Follow the instructions in section 4.3 on the process of setting up and using the spark environment.

Use Python code. We provide python notebooks for all the chapters on GitHub. Refer to this page `http://bit.ly/3r7cV4s` for a table with the links to the notebooks. Like R notebooks, you should be able to repeat all notebooks in your local machine except for Chapter 4 with reasons stated above. An easy way to repeat the notebook is to import and run in Google Colab. To use Colab, you only need to log in to your Google account in Chrome Browser. To load the notebook to your colab, you can do any of the following:

- Click the "Open in Colab" icon on the top of each linked notebook using the Chrome Brower. It should load the notebook and open it in your Colab.

- In your Colab, choose File -> Upload notebook -> GitHub. Copy-paste the notebook's link in the box, search, and select the notebook to load it. For example, you can load the python notebook for data preprocessing like this:

Examples	Recent	Google Drive	GitHub	Upload

Enter a GitHub URL or search by organization or user   ☐ Include private repos

https://github.com/happyrabbit/IntroDataScience/blob/master/Python/DataPreprocessing.ipynb   🔍

Repository: ☑
happyrabbit/IntroDataScience ▾

Branch: ☑
master ▾

Path

○ Python/DataPreprocessing.ipynb   🔍 ☑

○ Python/DataWrangling.ipynb   🔍 ☑

○ Python/FFNN.ipynb   🔍 ☑

○ Python/LoadDatasetSpark.ipynb   🔍 ☑

CANCEL

To repeat the code for big data, like running R notebook, you need to set up Spark in Databricks. Follow the instructions in section 4.3 on the process of setting up and using the spark environment. Then, run the "Create Spark Data" notebook to create Spark data frames. After that, you can run the pyspark notebook to learn how to use pyspark.

Complementary Reading

If you are new to R, we recommend R for Marketing Research and Analytics by Chris Chapman and Elea McDonnell Feit. The book is practical and provides repeatable R code. Part I & II of the book cover basics of R programming and foundational statistics. It is an excellent book on marketing analytics.

If you are new to Python, we recommend the Python version of the book mentioned above, Python for Marketing Research and Analytics by Jason Schwarz, Chris Chapman, and Elea McDonnell Feit.

If you want to dive deeper into some of the book's topics, there are many places to learn more.

- For machine learning, Python Machine Learning 3rd Edition by Raschka and Mirjalili is a good book on implementing machine learning in Python. Apply Predictive Modeling by Kuhn and Johnston is an applied, practitioner-friendly textbook using R package caret.

- For statistics models in R, a recommended book is An Introduction to Statistical Learning (ISL) by James, Witten, Hastie, and Tibshirani. A more advanced treatment of the topics in ISL is The Elements of Statistical Learning by Friedman, Tibshirani, and Hastie.

About the Authors

Hui Lin is a Lead Quantitative Researcher at Shopify. She holds an MS and Ph.D. in statistics from Iowa State University. Hui has worked across various industries, including traditional and high-tech, early-stage startups, and large corporations. She worked as a Marketing Data Scientist at DuPont; the first data hire at Netlify to build a data science team, and a Quantitative UX Researcher at Google. Aside from work, she is an active blogger on her website, https://scientistcafe.com/, where she shares her thoughts on data science. She was elected as the 2023 Chair of the American Statistical Association's Statistics in Marketing Section to advance statistical research and applications in marketing.

Ming Li is a Director of Data Science at PetSmart and an Adjunct Instructor of the University of Washington. He was the Chair of Quality & Productivity Section of the American Statistical Association for 2017. He was a Research Science Manager at Amazon, a Data Scientist at Walmart and a Statistical Leader at General Electric Global Research Center. He obtained his Ph.D. in Statistics from Iowa State University at 2010. With deep statistics background and a few years' experience in data science, he has trained and mentored numerous junior data scientists with different backgrounds such as statisticians, programmers, software developers, and business analysts. He was also an instructor of Amazon's internal Machine Learning University and was one of the key founding members of Walmart's Analytics Rotational Program.

Acknowledgment

We express our gratitude to everyone who made this book possible. First and foremost, we thank Hui's colleagues in the data science community at Google for their feedback on portions of the book. We are deeply grateful to Di Li, Xia Li, Richard Xu, Wenhao Hu, Chloe Ye, Yunchuan Kong, and Chris Chapman for their encouragement and comments.

We also thank the staff and editors at CRC for their developmental editing, copyediting, and proofreading. We are especially grateful to David Grubbs and Curtis Hill for their invaluable support and guidance.

We thank Alex Shum for his extensive feedback and comments, which were instrumental in shaping the final manuscript.

We acknowledge our families for their support and understanding during the editing process.

We thank you all for your contributions to this book. Without you, it would not have been possible.

1

Introduction

Data science is a rapidly evolving field. This chapter will explore various aspects of data science. We will discuss the various career paths and skills needed for data science, as well as the structure of a data science team. We will focus on the demand for data science from a business and industrial perspective. We hope this chapter will provide a complementary perspective to other data sciences books.

1.1 A Brief History of Data Science

Interest in data science-related careers is witnessing unprecedented growth and has seen a surge in popularity over the last few years. Data scientists come from a variety of backgrounds and disciplines, making it difficult to provide a concise answer when asked what data science is all about. Data science is a widely discussed topic, yet few can accurately define it.

Media has been hyping about "Data Science," "Big Data," and "Artificial Intelligence" over the past few years. There is an amusing statement from the internet:

"When you're fundraising, it's AI. When you're hiring, it's ML. When you're implementing, it's logistic regression."

For outsiders, data science is the magic that can extract useful information from data. Everyone is familiar with the concept of big data. Data science trainees must now possess the skills to manage

DOI: 10.1201/9781351132916-1

large data sets. These skills may include Hadoop, a system that uses Map/Reduce to process large data sets distributed across a cluster of computers or Spark, a system that builds on top of Hadoop to speed up the process by loading massive datasets into shared memory (RAM) across clusters with an additional suite of machine learning functions for big data.

The new skills are essential for dealing with large data sets beyond a single computer's memory or hard disk and the large-scale cluster computing. However, they are not necessary for deriving meaningful insights from data.

A lot of data means more sophisticated tinkering with computers, especially a cluster of computers. The computing and programming skills to handle big data were the biggest hurdle for traditional analysis practitioners to be a successful data scientist. However, this barrier has been significantly lowered thanks to the cloud computing revolution, as discussed in Chapter 2. After all, it isn't the size of the data that's important, but what you do with it. You may be feeling a mix of skepticism and confusion. We understand; we had the same reaction.

To declutter, let's start with a brief history of data science. If you search on Google Trends, which shows search keyword information over time, *you'll find that* the term "data science" dates back further than 2004. Media coverage may give the impression that machine learning algorithms are a recent invention and that there was no "big" data before Google. However, this is not true. While there are new and exciting developments in data science, many of the techniques we use are based on decades of work by statisticians, computer scientists, mathematicians, and scientists from a variety of other fields.

In the early 19th century, Legendre and Gauss came up with the least squares method for linear regression. At the time, it was mainly used by physicists to fit their data. Nowadays, anyone can build linear regression models using spreadsheet with just a little bit of self-guided online training.

In 1936, Fisher came up with linear discriminant analysis. In the 1940s, logistic regression became a widely used model. Then, in the 1970s, Nelder and Wedderburn formulated the "generalized linear mode (GLM)" which:

"generalized linear regression by allowing the linear model to be related to the response variable via a link function and by allowing the magnitude of the variance of each measurement to be a function of its predicted value." [from Wikipedia]

By the end of the 1970s, a variety of analytical models existed, most of them were linear due to the limited computing power available at the time. Non-linear models weren't able to be fitted until the 1980s.

In 1984, Breiman introduced the Classification and Regression Tree (CART), one of the oldest and most widely used classification and regression techniques (Breiman et al., 1984).

After that, Ross Quinlan developed tree algorithms such as ID3, C4.5, and C5.0. In the 1990s, ensemble techniques, which combine the predictions of many models, began to emerge. Bagging is a general approach that uses bootstrapping in conjunction with regression or classification models to construct an ensemble. Based on the ensemble idea, Breiman came up with the random forest model in 2001 (Breiman, 2001a). In the same year, Leo Breiman published a paper "Statistical Modeling: The Two Cultures" (Breiman, 2001b), in which he identified two cultures in the use of statistical modeling to extract information from data:

(1) Data is from a given stochastic data model
(2) Data mechanism is unknown and people approach the data using algorithmic model

Most of the classical statistical models are the first type of stochastic data model. Black-box models, such as random forest, **Gradient Boosting Machine (GBM)**, and deep learning, are algorithmic models. As Breiman pointed out, algorithmic models can be used on large, complex data as a more accurate and informative alternative to stochastic modeling on smaller datasets. These algorithms have developed rapidly with much-expanded applications in fields outside of traditional statistics which is one of the most important reasons why statisticians are not in the mainstream of today's data science, both in theory and practice.

Python is overtaking R as the most popular language in data science, mainly due to the backgrounds of many data scientists. Since 2000, the approaches to getting information out of data have shifted from traditional statistical models to a more diverse toolbox that includes machine learning and deep learning models. To help readers who are traditional data practitioners, we provide both R and Python codes.

What is the driving force behind the shifting trend? John Tukey identified four forces driving data analysis (there was no "data science" when this was written in 1962):

1. The formal theories of math and statistics
2. Acceleration of developments in computers and display devices
3. The challenge, in many fields, of more and ever larger bodies of data
4. The emphasis on quantification in an ever-wider variety of disciplines

Tukey's 1962 list is surprisingly modern, even when viewed in today's context. People often develop theories way before they find potential applications. Over the past 50 years, statisticians, mathematicians, and computer scientists have laid the theoretical groundwork for the construction of "data science" as we know it today.

The development of computers has enabled us to apply the algorithmic models (which can be very computationally expensive) and deliver results in a friendly and intuitive way. The transition to the internet and the internet of things has generated vast amounts of commercial data. Industries have also recognized the value of exploiting this data. Data science seems sure to be a significant preoccupation of commercial life in the coming decades. All the four forces John identified exist today and have been driving data science.

The applications have been expanding fast, benefiting from the increasing availability of digitized information and the ability to distribute it through the internet. Today, people apply data science in a variety of fields, such as business, health, biology, social science, politics, etc. But what is data science today?

1.2 Data Science Role and Skill Tracks

There is a widely diffused Chinese parable about a group of blind men conceptualizing what the elephant is like by touching it. The first person, whose hand landed on the trunk, said: "This being is like a thick snake." For another one whose hand reached its ear, it seemed like a fan. Another person whose hand was upon its leg said the elephant is a pillar-like tree trunk. The blind man who placed his hand upon its side said: "elephant is a wall." Another who felt its tail described it as a rope. The last felt its tusk, stating the elephant is hard, smooth like a spear.

Data science is the elephant. With the data science hype picking upstream, many professionals changed their titles to be "Data Scientist" without any necessary qualifications. Today's data scientists have vastly different backgrounds, yet each conceptualizes the elephant based on his/her professional training and application area. And to make matters worse, most of us are not even fully aware of our conceptualizations, much less the uniqueness of the experience from which they are derived.

> "We don't see things as they are, we see them as we are. [by Anais Nin]"

So, the answer to the question "what is data science?" depends on who you are talking to. Data science has three main skill tracks (figure 1.1): engineering, analysis, and modeling/inference (and yes, the order matters!).

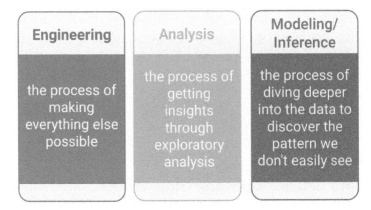

FIGURE 1.1
Three tracks of data science

There are some representative skills in each track. Different tracks and combinations of tracks will define different roles in data science.[1]

When people talk about all the machine learning and artificial intelligence algorithms, they often overlook the critical data engineering part that makes everything possible. Data engineering is the unseen iceberg under the water surface. Does your company need a data scientist? You are not ready for a data scientist if you don't have a data engineer yet. You need to have the ability to get data before making sense of it. If you only deal with small datasets with formatted data, you may be able to get by with plain text files such as CSV (i.e., comma-separated values) or even spreadsheet. As the data increasing in volume, variety, and velocity, data engineering becomes a sophisticated discipline in its own right.

1.2.1 Engineering

Data engineering is the foundation that makes everything else possible (figure 1.2). It mainly involves building data infrastructures

---

[1]This is based on "Industry recommendations for academic data science programs: `https://github.com/brohrer/academic_advisory`". It is a collection of thoughts of different data scientist across industries about what a data scientist does, and what differentiates an exceptional data scientist.

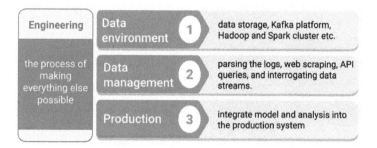

FIGURE 1.2
Engineering track

and pipelines. In the past when data was stored on local servers, computers, or other devices, constructing the data infrastructure was a major IT project. This included software, hardware for servers to store the data, and the ETL (extract, transform, and load) process.

With the advent of cloud computing, the new standard for storing and computing data is on the cloud. Data engineering today is essentially software engineering with data flow as the primary focus. The fundamental element for automation is maintaining the data pipeline through modular, well-commented code, and version control.

(1) Data environment

Designing and setting up the entire environment to support data science workflow is the prerequisite for data science projects. It may include setting up storage in the cloud, Kafka platform, Hadoop and Spark clusters, etc. Each company has a unique data condition and need. The data environment will be different depending on the size of the data, update frequency, the complexity of analytics, compatibility with the back-end infrastructure, and (of course) budget.

(2) Data management

Automated data collection is a common task that includes parsing the logs (depending on the stage of the company and the type of industry you are in), web scraping, API queries, and

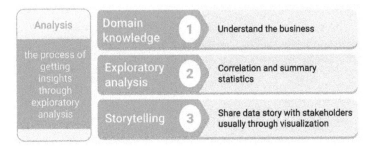

FIGURE 1.3
Analysis track

interrogating data streams. Data management includes constructing data schema to support analytics and modeling needs, and ensuring data is correct, standardized, and documented.

(3) Production

If you want to integrate the model or analysis into the production system, you have to automate all data handling steps. It involves the whole pipeline from data access, preprocessing, modeling to final deployment. It is necessary to make the system work smoothly with all existing software stacks. So, it requires monitoring the system through some robust measures, such as rigorous error handling, fault tolerance, and graceful degradation to make sure the system is running smoothly and users are happy.

1.2.2 Analysis

Analysis turns raw information into insights into a fast and often exploratory way. In general, an analyst needs to have decent domain knowledge, do exploratory analysis efficiently, and present the results using storytelling (figure 1.3).

(1) Domain knowledge

Domain knowledge is the understanding of the organization or industry where you apply data science. You can't make sense of data without context. Some questions about the context are

- What are the critical metrics for this kind of business?

- What are the business questions?
- What type of data do they have, and what does the data represent?
- How to translate a business need to a data problem?
- What has been tried before, and with what results?
- What are the accuracy-cost-time trade-offs?
- How can things fail?
- What are other factors not accounted for?
- What are the reasonable assumptions, and what are faulty?

Domain knowledge helps you to deliver the results in an audience-friendly way with the right solution to the right problem.

(2)   Exploratory analysis

This type of analysis is about exploration and discovery. Rigorous conclusions are not the primary driver, which means the goal is to get insights driven by correlation, not causation. The latter one requires more advanced statistical skills and hence more time and resource expensive. Instead, this role will help your team look at as much data as possible so that the decision-makers can get a sense of what's worth further pursuing. It often involves different ways to slice and aggregate data. An important thing to note here is that you should be careful not to get a conclusion beyond the data. You don't need to write production-level robust codes to perform well in this role.

(3)   Storytelling

Storytelling with data is critical to deliver insights and drive better decision making. It is the art of telling people what the numbers signify. It usually requires data summarization, aggregation, and visualization. It is crucial to answering the following questions before you begin down the path of creating a data story.

- Who is your audience?
- What do you want your audience to know or do?
- How can you use data to help make your point?

A business-friendly report or an interactive dashboard is the typical outcome of the analysis.

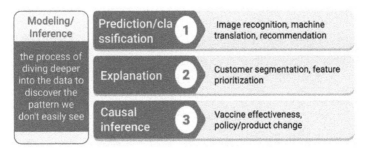

FIGURE 1.4
Modeling/inference track

1.2.3   Modeling/Inference

Modeling/inference is a process that dives deeper into the data to discover patterns that are not easily seen. It is often misunderstood. When people think of data science, they may immediately think of complex machine learning models. Despite the overrepresentation of machine learning in the public's mind, the truth is that you don't have to use machine learning to be a data scientist. Even data scientists who use machine learning in their work spend less than 20% of their time working on machine learning. They spend most of their time communicating with different stakeholders and collecting and cleaning data.

This track mainly focuses on three problems: (1) prediction, (2) explanation, and (3) causal inference (figure 1.4)).

Prediction focuses on predicting based on what has happened, and understanding each variable's role is not a concern. Many black-box models, such as ensemble methods and deep learning, are often used to make a prediction. Examples of problems are image recognition, machine translation, and recommendation. Despite the remarkable success of many deep-learning models, they operate almost entirely in an associational mode. As Judea Pearl pointed out in his book "The book of why" (Pearl and Mackenzie, 2019), complex black-box algorithms like AlphaGo "do not really know why it works, only that it does." Judea Pearl came up with a Ladder of Causation (Pearl and Mackenzie, 2019) with three levels: 1) association, 2) intervention, and 3) counterfactuals. According to this framework, prediction problems are on the first level.

The next level of the ladder, intervention, requires model interpretability. Questions on this level involve not just seeing but changing. The question pattern is like, "what happens if I do ...?" For example, product managers often need to prioritize a list of features by user preference. They need to know what happens if we build feature a instead of feature b. Choice modeling is a standard method for this problem which allows you to explain the drivers behind users' decisions on using/purchasing a product. Another way to directly study the result of an intervention is through experiments. Tech companies constantly perform A/B tests to examine what happens if they make some product change.

Causal inference is on the third level, which is counterfactual. When an experiment is not possible, and the cost of a wrong decision is too high, you need to use the existing data to answer a counterfactual question: "what would have happened if I had taken a different approach?" For example, if you are a policymaker who wants to find a way to reduce the divorce rate in the United States, you see from the data the southern states have a higher divorce rate and a lower median age at marriage. You may assume that getting married when you are young leads to a higher chance of divorce. But it is impossible to experiment by randomly selecting a group of people and asking them to get married earlier and the other group to get married later as a control. In this case, you must find a way to match samples or create a balanced pseudo-population. This type of problem is out of the scope of this book. For a non-technical introduction to causal inference, readers can refer to "The book of why" (Pearl and Mackenzie, 2019).

If we look at this track through the lens of the technical methods used, there are three types.

(1) Supervised learning

In supervised learning, each sample corresponds to a response measurement. There are two flavors of supervised learning: regression and classification. In regression, the response is a real number, such as the total net sales in 2017 for a company or the yield of wheat next year for a state. The goal for regression is to approximate the response measurement. In classification, the response is a class label, such as a dichotomous response of yes/no. The response can also have more than two categories, such as four segments of customers. A supervised learning model is a function that maps some input variables (X) with corresponding parameters (beta) to a response (y). The modeling process is to adjust the value of parameters to make the mapping fit the given response. In other words, it is to minimize the discrepancy between given responses and the model output. When the response y is a real value number, it is intuitive to define discrepancy as the squared difference between model output and the response. When y is categorical, there are other ways to measure the difference, such as the area under the receiver operating characteristic curve (i.e., AUC) or information gain.

(2) Unsupervised learning

In unsupervised learning, there is no response variable. For a long time, the machine learning community overlooked unsupervised learning except clustering. Moreover, many researchers thought that clustering was the only form of unsupervised learning. One reason is that it is hard to define the goal of unsupervised learning explicitly. Unsupervised learning can be used to do the following:

- Identify a good internal representation or pattern of the input that is useful for subsequent supervised or reinforcement learning, such as finding clusters;

- It is a dimension reduction tool that provides compact, low dimensional representations of the input, such as factor analysis.

- Provide a reduced number of uncorrelated learned features from original variables, such as principal component regression.

(3) Customized model development

In most cases, after a business problem is fully translated into a data science problem, a data scientist needs to use out of the box algorithms to solve the problem with the right data. But in some situations, there isn't enough data to use any machine learning model, or the question doesn't fit neatly in the specifications of existing tools, or the model needs to incorporate some prior domain knowledge. A data scientist may need to develop new models to accommodate the subtleties of the problem at hand. For example, people may use Bayesian models to include domain knowledge as the modeling process's prior distribution.

Here is a list of questions that can help you decide the type of technique to use:

- Is your data labeled? It is straightforward since supervised learning needs labeled data.

- Do you want to deploy your model at scale? There is a fundamental difference between building and deploying models. It is like the difference between making bread and making a bread machine. One is a baker who will mix and bake ingredients according to recipes to make a variety of bread. One is a machine builder who builds a machine to automate the process and produce bread at scale.

- Is your data easy to collect? One of the major sources of cost in deploying machine learning is collecting, preparing, and cleaning the data. Because model maintenance includes continuously collecting data to keep the model updated. If the data collection process requires too much human labor, the maintenance cost can be too high.

- Does your problem have a unique context? If so, you may not be able to find any off-the-shelf method that can directly apply to your question and need to customize the model.

What others?

There are some common skills to have, regardless of the role people have in data science.

- **Data preprocessing: the process nobody wants to go through yet nobody can avoid**

No matter what role you hold in the data science team, you will have to do some data cleaning, which tends to be the least enjoyable part of anyone's job. Data preprocessing is the process of converting raw data into clean data.

(1)   Data preprocessing for data engineer

Getting data from different sources and dumping them into a data lake. A data lake is a storage repository that stores a vast amount of raw data in its native format, including XML, JSON, CSV, Parquet, etc. It is a data cesspool rather than a data lake. The data engineer's job is to get a clean schema out of the data lake by transforming and formatting the data. Some common problems to resolve are

- Enforce new tables' schema to be the desired one
- Repair broken records in newly inserted data
- Aggregate the data to form the tables with a proper granularity

(2)   Data preprocessing for data analyst and scientist

Not just for a data engineer, preprocessing also occupies a large portion of data analyst and scientist's working hours. A facility and a willingness to do these tasks are a prerequisite for a good data scientist. If you are lucky as a data scientist, you may end up spending 50% of your time doing this. If you are like most of us, you will probably spend over 80% of your working hours wrangling data.

The data a data scientist gets can still be very rough even if it is from a nice and clean database that a data engineer sets up. For example, dates and times are notorious for having many representations and time zone ambiguity. You may also get market survey responses from your clients in an excel file where the table title could be multi-line, or the format does not meet the requirements, such as using 50% to represent the percentage rather than 0.5. In many cases, you need to set the data to be the right format before moving on to analysis.

Even the data is in the right format. There are other issues to solve before or during analysis and modeling. For example, variables can have missing values. Knowledge about the data collection process and what it will be used for is necessary to decide a way to handle the missing. Also, different models have different requirements for the data. For example, some models may require a consistent scale; some may be susceptible to outliers or collinearity; some may not be able to handle categorical variables, and so on. The modeler has to preprocess the data to make it proper for the specific model.

Most of the people in data science today focus on one of the tracks. A small number of people are experts on two tracks.

1.3 What Kind of Questions Can Data Science Solve?

1.3.1 Prerequisites

Data science is not a cure-all, and there are issues it cannot address. It is best to make a decision as soon as possible in the analytical process. Above all, we must be honest and transparent with customers, clients, and stakeholders when we believe data analytics cannot answer their questions after a thorough evaluation of the request, data availability, computing resources, and modeling details. Frequently, we can suggest an alternative. It is essential to "negotiate" with others what data science can do specifically; simply answering "we cannot do what you want" will end the collaboration. Now let's see what kind of questions data science can solve:

1. The question needs to be specific enough

   Let us look at the two examples below:

   - Question 1: How can I increase product sales?
   - Question 2: Is the new promotional tool introduced at the beginning of this year boosting the annual sales of P1197 in Iowa and Wisconsin? (P1197 is a corn seed product)

   It is easy to see the difference between the two questions. Question 1 is grammatically correct, but it is not proper for data analysis to answer. Why? It is too general. What is the response variable here? Product sales? Which product? Is it annual sales or monthly sales? What are the candidate predictors? We nearly can't get any useful information from the questions.

In contrast, question 2 is much more specific. From the analysis point of view, the response variable is clearly "annual sales of P1197 in Iowa and Wisconsin." Even if we don't know all the predictors, the variable of interest is "the new promotional tool introduced early this year." We want to study the impact of the promotion on sales. We can start there and figure out other variables that need to be included in the model.

As a data scientist, we may start with general questions from customers, clients, or stakeholders and eventually get to more specific and data science solvable questions with a series of communication, evaluation, and negotiation. Effective communication and in-depth knowledge about the business problems are essential to converting a general business question into a solvable analytical problem. Domain knowledge helps data scientists communicate using the language other people can understand and obtain the required context.

Defining the question and variables involved won't guarantee that we can answer it. For example, we could encounter this situation with a well-defined supply chain problem. The client may ask us to estimate the stock needed for a product in a particular area. Why can't this question be answered? We can try to fit various models such as a multivariate adaptive regression spline (MARS) model and find a reasonable solution from a modeling perspective. But it can turn out later that the client's data is an estimated value, not the actual observation. There is no good way for data science to solve the problem with the desired accuracy with inaccurate data.

2. You need to have accurate and relevant data

One cannot make a silk purse out of a sow's ear. Data scientists relevant and accurate data. The supply problem mentioned above is a case in point. There was relevant data, but not sound. All the later analytics based on that data was a building on sand.

Of course, data nearly almost have noise, but it has to be in a certain range. Generally speaking, the accuracy requirement for the independent variables of interest and response variable is higher than others. For the above question 2, it is variables related to the "new promotion" and "sales of P1197."

The data has to be helpful for the question. If we want to predict which product consumers are most likely to buy in the next three months, we need to have historical purchasing data: the last buying time, the amount of invoice, coupons, etc. Information about customers' credit card numbers, ID numbers, and email addresses will not help much.

Often, the data quality is more important than the quantity, but you cannot completely overlook quantity. Suppose you can guarantee data quality, even then the more data, the better. If we have a specific and reasonable question with sound and relevant data, then congratulations, we can start playing data science!

1.3.2 Problem Type

Many of the data science books classify various models from a technical point of view. Such as supervised vs. unsupervised models, linear vs. nonlinear models, parametric models vs. non-parametric models, and so on. Here we will continue on a "problem-oriented" track. We first introduce different groups of real-world problems and then present which models can answer the corresponding category of questions.

1. Description

The primary analytic problem is to summarize and explore a data set with descriptive statistics (mean, standard deviation, and so forth) and visualization methods. It is the most straightforward problem and yet the most crucial and common one. We will need to describe and explore the dataset before moving on to a more complex analysis. For problems such as customer segmentation, after we cluster the sample, the next step is to figure out each class's profile by comparing the descriptive statistics of various variables. Questions of this kind are

- What is the annual income distribution?
- Are there any outliers?
- What are the mean active days of different accounts?

Data description is often used to check data, find the appropriate data preprocessing method, and demonstrate the model results.

2. Comparison

The first common modeling problem is to compare different groups. Is A better in some way than B? Or more comparisons: Is there any difference among A, B, and C in a particular aspect? Here are some examples:

- Are males more inclined to buy our products than females?
- Are there any differences in customer satisfaction in different business districts?
- Do soybean carrying a particular gene have higher oil content?

For those problems, it usually starts with some summary statistics and visualization by groups. After a preliminary visualization, you can test the differences between the treatment and control groups statistically. The commonly used statistical tests are chi-square test, t-test, and ANOVA. There are also methods using Bayesian methods. In the biology industry, such as new drug development, crop breeding, fixed/random/mixed effect models are standard techniques.

3. Clustering

Clustering is a widespread problem, and it can answer questions like:

- How many reasonable customer segments are there based on historical purchase patterns?
- How are the customer segments different from each other?

Please note that clustering is unsupervised learning; there are no response variables. The most common clustering algorithms include K-Means and Hierarchical Clustering.

4. Classification

For classification problems, there are one or more label columns to define the ground truth of classes. We use other features of the training dataset as explanatory variables for model training. We can use the trained classifier to predict the labels of a new observation. Here are some example questions:

- Will this customer likely to buy our product?
- Is the borrower going to pay us back?
- Is it spam email or not?

There are hundreds of different classifiers. In practice, we do not need to try all the models but several models that perform well generally. For example, the random forest algorithm is usually used as the baseline model to set model performance expectations.

5. Regression

In general, regression deals with a question like "how much is it?" and return a numerical answer. It is necessary to coerce the model results to be 0 or round it to the nearest integer in some cases. It is still the most common problem in the data science world.

- What will be the temperature tomorrow?
- What is the projected net income for the next season?
- How much inventory should we have?

6. Optimization

Optimization is another common type of problems in data science to find an optimal solution by tuning a few tune-able variables with other non-controllable environmental variables. It is an expansion of comparison problem and can solve problems such as:

- What is the best route to deliver the packages?
- What is the optimal advertisement strategy to promote a new product?

1.4 Structure of Data Science Team

A vast amount of data has become available and readily accessible for analysis in many companies across different business sectors during the past decade. The size, complexity, and speed of increment of data suddenly beyond the traditional scope of statistical analysis or business intelligence (i.e., BI) reporting. To leverage the big data collected, do you need an internal data science team to be a core competency, or can you outsource it? The answer depends on the problems you want to solve using data. If they are critical to the business, you can't afford to outsource it. Also, each company has its business context, and it needs new kinds of data as the business grows and uses the results in novel ways. Being a data-driven organization requires cross-organization commitments to identify what data each department needs to collect, establish the infrastructure and process for collecting and maintaining that data, and standardize how to deliver analytical results. Unfortunately, it is unlikely that an off-the-shelf solution will be flexible enough to adapt to the specific business context. In general, most of the companies establish their data science team.

Where should the data science team fit? In general, the data science team is organized in three ways.

(1) A standalone team

Data science is an autonomous unit parallel to the other organizations (such as engineering, product, etc.). The head of data science reports directly to senior leadership, ideally to the CEO or at least someone who understands data strategy and is willing to invest and give it what it needs. The advantages of this type of data organization are

- The data science team has autonomy and is well-positioned to tackle whatever problems it deems important to the company.
- It is advantageous for people in the data science team to share knowledge and grow professionally.

- It provides a clear career path for data science professionals and shows the company treats data as a first-class asset. So, it tends to attract and retain top talent people.

The biggest concern of this type of organization is the risk of marginalization. Data science only has value if data drives action, which requires collaboration among data scientists, engineers, product managers, and other business stakeholders across the organization. Suppose you have a standalone data science team. It is critical to choose a data science leader who is knowledgeable about the applications of data science in different areas and has strong inter-discipline communication skills. The head of data science needs to build a strong collaboration with other departments.

As companies grow, each department prefers to be self-sufficient and tries to hire its data personal under different titles even when they can get support from the standalone data science team. That is why it is unlikely for an already mature company to have a standalone data science team. If you start your data science team in the early stage as a startup, it is important that the CEO sets a clear vision from the beginning and sends out a strong message to the whole company about accessing data support.

(2)   An embedded model

There is still a head of data science, but his/her role is mostly a hiring manager and coach, and he/she may report to a senior manager in the engineering department. The data science team brings in talented people and farms them out to the rest of the company. In other words, it gives up autonomy to ensure utility. The advantages are

- Data science is closer to its applications.
- There is still a data science group, so it is easy to share knowledge.
- It has high flexibility to allocate data science resources across the company.

However, there are also concerns.

- It brings difficulty to the management since the designated team's lead is not responsible for data science professionals' growth and happiness. In contrast, data science managers are not directly vested in their work.
- Data scientists are second-class citizens everywhere, and it is hard to attract and retain top talent.

(3) Integrated team

There is no data science team. Each team hires its data science people. For example, a marketing analytics group consists of a data engineer, data analyst, and data scientist. The team leader is a marketing manager who has an analytical mind and in-depth business knowledge. The advantages are apparent.

- Data science resource aligns with the organization very well.
- Data science professionals are first-class members and valued in their team. The manager is responsible for data science professionals' growth and happiness.
- The insights from the data are quickly put into action.

It works well in the short term for both the company and the data science hires. However, there are also many concerns.

- It sacrifices data science hires' professional growth since they work in silos and specialize in a specific application. It is also difficult to share knowledge across different applied areas.
- It is harder to move people around since they are highly associated with a specific organization's specific function.
- There is no career path for data science people, and it is difficult to retain talent.

There is no universal answer to the best way to organize the data science team. It depends on the answer to many other questions. How important do you think the data science team is for your company? What is the stage of your company when you start to build a data science team? Are you a startup or a relatively mature company? How valuable it is to use data to tell the truth, how dangerous it is to use data to affirm existing opinions.

Data science has its skillset, workflow, tooling, integration process, and culture. If it is critical to your organization, it is best not to bury it under any part of the organization. Otherwise, data science will only serve the need for a specific branch. No matter which way you choose, be aware of both sides of the coin. If you are looking for a data science position, it is crucial to know where the data science team fits.

1.5 Data Science Roles

As companies learn about using data to help with the business, there is a continuous specialization of different data science roles. As a result, the old "data scientist" title is fading, and some other data science job titles are emerging. In the past, misunderstanding data science's fundamental work led to confusing job postings and frustrations for stakeholders and data scientists. Stakeholders were frustrated that they weren't getting what they expected, and data scientists were frustrated that the company didn't appreciate their talent.

On the one hand, the competitive hiring market has pushed organizations to have a streamlined and transparent interview process. They must clarify the role and responsibilities, tool usage, and daily work for the candidates to understand what the role entails. Role clarity is critical for building a career path and retaining data science talents. As a result, we are glad to see the job definition **within an organization** has improved dramatically.

On the other hand, however, there is **title inconsistency across different companies or industries**, especially for the analytical roles (i.e., data analysts and data scientists). An analyst at one company may be close to a data scientist at another company.

The following table shows a list of data science job titles. Some are relatively new, and others have been around for some time but are now well-defined. In the rest of this section, we will illustrate different data science roles, backgrounds, and required skills in

general. The title and profile combination in the following text may
not represent the truth of a particular company. You may find the
description of a role under a different title.

Role	Skills
Data infrastructure engineer	Go, Python, AWS/Google Cloud/Azure, logstash, Kafka, and Hadoop
Data engineer	spark/scala, python, SQL, AWS/Google Cloud/Azure, Data modeling
BI engineer	Tableau/looker/Mode, etc., data visualization, SQL, Python
Data analyst	SQL, basic statistics, data visualization
Data scientist	R/Python, SQL, basic applied statistics, data visualization, experimental design
Research scientist	R/Python, advanced statistics, experimental design, ML, research background, publications, conference contributions, algorithms
Applied scientist	ML algorithm design, often with an expectation of fundamental software engineering skills
Machine learning engineer	More advanced software engineering skillset, algorithms, machine learning algorithm design, system design

The above table shows some data science roles and common
technical keywords in job descriptions. Those roles are different in
the following key aspects:

- How much business knowledge is required?
- Does it need to deploy code in the production environment?

	Business Knowledge	Data Frequency	Engineering Skill	Math/Stat	Production	(Un)Str Data
Data infrastructure engineer	Low	High	High	Low	Yes	Both
Data engineer	Low/Mid	High	High	Low	Yes	Both
BI engineer	High	Mid	Mid	Mid	Depends	Str
Data analyst	High	Mid	Low/Mid	Mid	No	Str
Data scientist	High	Mid	Low/Mid	High	Mostly No	Mostly Str
Research scientist	High	Mid	Low/Mid	High	No	Mostly Str
Applied scientist	High	Mid/High	Mid/High	Mid/High	Depends	Both
Machine Learning Engineer	Low	High	High	Mid	Yes	Both

FIGURE 1.5
Different roles in data science and the skill requirements

- How frequently is data updated?
- How much engineering skill is required?
- How much math/stat knowledge is needed?
- Does the role work with structured or unstructured data?

Data infrastructure engineers work at the beginning of the data pipeline. They are software engineers who work in the production system and usually handle high-frequency data. They are responsible for bringing data of different forms and formats and ensuring data comes in smoothly and correctly. They work directly with other engineers (for example, data engineers and backend engineers). They typically don't need to know the data's business context or how data scientists will use it. For example, integrate the company's services with AWS/GCP/Azure services and set up an Apache Kafka environment to stream the events.

People call a storage repository with vast raw data in its native format (XML, JSON, CSV, Parquet, etc.) a **data lake** (figure 1.6). As the number of data sources multiplies, having data scattered in various formats prevents the organization from using the data to help with business decisions or building products. That is when data engineers come to help.

Data engineers transform, clean, and organize the data from the data lake. They commonly design schemas, store data in queryable forms, and build and maintain data warehouses. People call this cleaner and better-organized database **data mart** (figure 1.6) which contains a subset of data for business needs. They use technologies like Hadoop/Spark and SQL. Since the database is

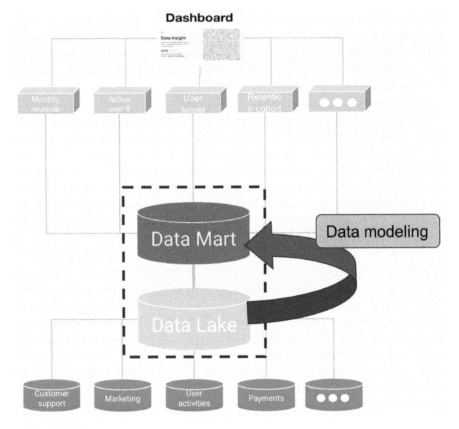

FIGURE 1.6
Data lake (a focused version of a data warehouse that contains
a subset of data for business needs) and data mart (a storage
repository that cheaply stores a vast amount of raw data in its
native format (XML, JSON, CSV, Parquet, etc.))

for non-engineers, data engineers need to know more about the
business and how analytical personnel uses the data. Some may
have a basic understanding of machine learning to deploy models
developed by data/research scientists.

Business intelligence (BI) engineers and data analysts are close
to the business, so they need to know the business context well. The
critical difference is that BI engineers build automated dashboards,
so they are engineers. They are usually experts in SQL and have
the engineering skill to write production-level code to construct the
later steam data pipeline and automate their work. Data analysts

are technical but not engineers. They analyze ad hoc data and deliver the results through presentations. The data is, most of the time, structured. They need to know coding basics (SQL or R/Python), but they rarely need to write production-level code. This role was mixed with "data scientist" by many companies but is now much better refined in mature companies.

The most significant difference between a data analyst and a data scientist is the requirement of mathematics and statistics. Most data scientists have a quantitative background and do A/B experiments and sometimes machine learning models. Data analysts usually don't need a quantitative background or an advanced degree. The analytics they do are primarily descriptive with visualizations. They mainly handle structured and ad hoc data.

Research scientists are experts who have a research background. They do rigorous analysis and make causal inferences by framing experiments and developing hypotheses, and proving whether they are true or not. They are researchers that can create new models and publish peer-reviewed papers. Most of the small/mid companies don't have this role.

Applied scientist is the role that aims to fill the gap between data/research scientists and data engineers. They have a decent scientific background but are also experts in applying their knowledge and implementing solutions at scale. They have a different focus than research scientists. Instead of scientific discovery, they focus on real-life applications. They usually need to pass a coding bar.

In the past, some data scientist roles encapsulated statistics, machine learning, and algorithmic knowledge, including taking models from proof of concept to production. However, more recently, some of these responsibilities are now more common in another role: machine learning engineer. Often larger companies may distinguish between data scientists and machine learning engineer roles. Machine learning engineer roles will deal more with the algorithmic and machine learning side and strongly emphasize software engineering. In contrast, data scientist roles will emphasize analytics (as with data analysts) and statistics, such as significance testing and causal inference.

2

Soft Skills for Data Scientists

There are many university courses, online self-learning modules, and excellent books that teach technical skills. However, there are much fewer resources discussing the soft skills for data scientists in detail. However, soft skills are essential for data scientists to succeed in their career, especially in the early stage. We want to introduce soft skills for data scientists before discussing technical components. This chapter also introduces the project cycle and some common pitfalls of data science projects in real life.

2.1 Comparison between Statistician and Data Scientist

Statistics as a scientific area can be traced back to 1749, and statistician as a career has been around for hundreds of years with well-established theory and application. Data scientist becomes an attractive career for only a few years, along with the fact that data size and variety beyond the traditional statistician's toolbox and the fast-growing of computation power. Statistician and data scientist have a lot in common, but there are also significant differences, as highlighted in figure 2.1.

Both statisticians and data scientists work closely with data. For typical traditional statisticians, the data set is usually well-formatted text files with numbers (i.e., numerical variables) and labels (i.e., categorical variables). The data's size is typically small enough to be loaded in a PC's memory or be saved in a PC's hard disk. Comparing to statisticians, data scientists need to deal with more varieties of data:

DOI: 10.1201/9781351132916-2

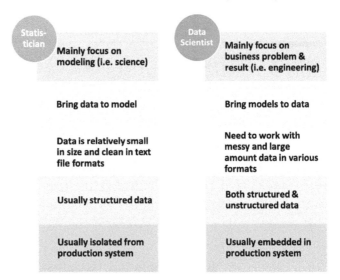

FIGURE 2.1
Comparison of statistician and data scientist

- well-formatted data stored in a database system with a size much larger than a PC's memory or hard-disk;
- a huge amount of verbatim text, voice, image, and video;
- real-time streaming data and other types of records.

One unique power of statistics is to make statistical inferences based on a small set of data. Statisticians, especially in academia, usually spend most of their time developing models and don't need to put too much effort into data cleaning. However, data becomes relatively abundant recently, and modeling is (often small) part of the overall effort. Due to open source communities' active development, fitting standard models are not too far away from button-pushing. Data scientists in industry instead spend a lot of time preprocessing and wrangling the data before feeding them to the model.

Unlike statisticians, data scientists often focus on delivering actionable results and sometimes need to deploy the model to the production system. The data available for model training can be too large to be processed in a single computer. From the entire problem-solving cycle, statisticians are usually not well integrated with the production system where data is obtained in real-time, while data scientists are more embedded in the production system and closer to the data generation procedures. In summary, statisticians focus more on modeling and usually bring data to models, while data scientists focus more on data and usually bring models to data.

2.2 Beyond Data and Analytics

Data scientists usually have a good sense of data and analytics, but data science projects are much more than that. A data science project may involve people with different roles, especially in a large company:

- the business owner or leader who identifies business problem and value;
- the data owner and computation resource/infrastructure owner from the IT department;
- a dedicated policy owner to make sure the data and model are under model governance, security and privacy guidelines and laws;
- a dedicated engineering team to implement, maintain and refresh the model;
- a program manager to ensure the data science project fits into the overall technical program development and to coordinate all involved parties to set periodical tasks so that the project meets the preset milestones and results;

The entire team usually will have multiple rounds of discussion of resource allocation among groups (i.e., who pay for the data science project) at the beginning of the project and during the project.

Effective communication and in-depth domain knowledge about the business problem are essential requirements for a successful data scientist. A data scientist may interact with people at various levels, from senior leaders who set the corporate strategies to front-line employees who do the daily work. A data scientist needs to have the capability to view the problem from 10,000 feet above the ground and down to the detail to the very bottom. To convert a business question into a data science problem, a data scientist needs to communicate using the language other people can understand and obtain the required information through formal and informal conversations.

In the entire data science project cycle, including defining, planning, developing, and implementing, every step needs to get a data scientist involved to ensure the whole team can correctly determine the business problem and reasonably evaluate the business value and success. Corporates are investing heavily in data science and machine learning, and there is a very high expectation of return for the investment.

However, it is easy to set an unrealistic goal and inflated estimation for a data science project's business impact. The team's data scientist should lead and navigate the discussions to ensure data and analytics, not wishful thinking, back the goal. Many data science projects often over-promise in business value and are too optimistic on the timeline to delivery. These projects eventually fail by not delivering the pre-set business impact within the promised timeline. As data scientists, we need to identify these issues early and communicate with the entire team to ensure the project has a realistic deliverable and timeline. The data scientist team also needs to work closely with data owners on different things. For example, identify a relevant internal and external data source, evaluate the data's quality and relevancy to the project, and work closely with the infrastructure team to understand the computation resources (i.e., hardware and software) availability. It is easy to create scalable computation resources through the cloud infrastructure for a data science project. However, you need to evaluate the dedicated computation resources' cost and make sure it fits the budget.

In summary, data science projects are much more than data and analytics. A successful project requires a data scientist to lead many aspects of the project.

2.3 Three Pillars of Knowledge

It is well known there are three pillars of essential knowledge for a successful data scientist.

(1) Analytics knowledge and toolsets

A successful data scientist needs to have a strong technical background in data mining, statistics, and machine learning. The in-depth understanding of modeling with insight about data enables a data scientist to convert a business problem to a data science problem. Many chapters of this book are focusing on analytics knowledge and toolsets.

(2) Domain knowledge and collaboration

A successful data scientist needs in-depth domain knowledge to understand the business problem well. For any data science project, the data scientist needs to collaborate with other team members. Communication and leadership skills are critical for data scientists during the entire project cycle, especially when there is only one scientist in the project. The scientist needs to decide the timeline and impact with uncertainty.

(3) (Big) data management and (new) IT skills

The last pillar is about computation environment and model implementation in a big data platform. It used to be the most difficult one for a data scientist with a statistics background (i.e., lack computer science knowledge or programming skills). The good news is that with the rise of the big data platform in the cloud, it is easier for a statistician to overcome this barrier. The "Big Data Cloud Platform" chapter of this book will describe this pillar in detail.

2.4 Data Science Project Cycle

A data science project has various stages. Many textbooks and blogs focus on one or two specific stages, and it is rare to see an

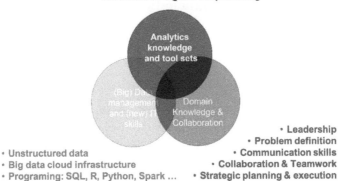

- **Data understanding and processing**
- **Statistical methods and problem solving**
- **Machine learning and deep learning**

- **Leadership**
- **Problem definition**
- **Communication skills**
- **Collaboration & Teamwork**
- **Strategic planning & execution**

- Unstructured data
- Big data cloud infrastructure
- Programing: SQL, R, Python, Spark ...

FIGURE 2.2
Three pillars of knowledge

end-to-end life cycle of a data science project. To get a good grasp of the end-to-end process requires years of real-world experience. Seeing a holistic picture of the whole cycle helps data scientists to better prepare for real-world applications. We will walk through the full project cycle in this section.

2.4.1 Types of Data Science Projects

People often use data science projects to describe any project that uses data to solve a business problem, including traditional business analytics, data visualization, or machine learning modeling. Here we limit our discussion of data science projects that involve data and some statistical or machine learning models and exclude basic analytics or visualization. The business problem itself gives us the flavor of the project. We can view data as the raw ingredient to start with, and the machine learning model makes the dish. The types of data used and the final model development define the different kinds of data science projects.

2.4.1.1 Offline and Online Data

There are offline and online data. Offline data are historical data stored in databases or data warehouses. With the development of data storage techniques, the cost to store a large amount of data is low. Offline data are versatile and rich in general (for example, websites may track and keep each user's mouse position, click and typing information while the user is visiting the website). The data is usually stored in a distributed system, and it can be extracted in batch to create features used in model training.

Online data are real-time information that flows to models to make automatic actions. Real-time data can frequently change (for example, the keywords a customer is searching for can change at any given time). Capturing and using real-time online data requires the integration of a machine learning model to the production infrastructure. It used to be a steep learning curve for data scientists not familiar with computer engineering, but the cloud infrastructure makes it much more manageable. Based on the offline and online data and model properties, we can separate data science projects into three different categories as described below.

2.4.1.2 Offline Training and Offline Application

This type of data science project is for a specific business problem that needs to be solved once or multiple times. But the dynamic and disruptive nature of this type of business problem requires substantial work every time. One example of such a project is "whether a brand-new business workflow is going to improve efficiency." In this case, we often use internal/external offline data and business insight to build models. The final results are delivered as a report to answer the specific business question. It is similar to the traditional business intelligence project but with more focus on data and models. Sometimes the data size and model complexity are beyond the capacity of a single computer. Then we need to use distributed storage and computation. Since the model uses historical data, and the output is a report, there is no need for real-time execution. Usually, there is no run-time constraint on the machine learning model unless the model runs beyond a reasonable time frame, such as a few days. We can call this type of data science project "offline training, offline application" project.

2.4.1.3   Offline Training and Online Application

Another type of data science project uses offline data for training and applies the trained model to real-time online data in the production environment. For example, we can use historical data to train a personalized advertisement recommendation model that provides a real-time ad recommendation. The model training uses historical offline data. The trained model then takes customers' online real-time data as input features and run the model in real-time to provide an automatic action. The model training is very similar to the "offline training, offline application" project. But to put the trained model into production, there are specific requirements. For example, as features used in the offline training have to be available online in real-time, the model's online run-time has to be short enough without impacting user experience. In most cases, data science projects in this category create continuous and scalable business value as the model could run millions of times a day. We will use this type of data science project to describe the typical data science project cycle from section 2.4.2 to section 2.4.5.

2.4.1.4   Online Training and Online Application

For some business problems, it is so dynamic that even yesterday's data is out of date. In this case, we can use online data to train the model and apply it in real-time. We call this type of data science project "online training, online application." This type of data science project requires high automation and low latency.

2.4.2   Problem Formulation and Project Planning Stage

A data-driven and fact-based planning stage is essential to ensure a successful data science project. With the recent big data and data science hype, there is a high demand for data science projects to create business value across different business sectors. Usually, the leaders of an organization are those who initiate the data science project proposals. This top-down style of data science projects typically have high visibility with some human and computation resources pre-allocated. However, it is crucial to understand the business problem first and align the goal across different teams, including:

(1) the business team, which may include members from the business operation team, business analytics, insight, and metrics reporting team;

(2) the technology team, which may include members from the database and data warehouse team, data engineering team, infrastructure team, core machine learning team, and software development team;

(3) the project, program, and product management team depending on the scope of the data science project.

To start the conversation, we can ask everyone in the team the following questions :

- What are the pain points in the current business operation?
- What data are available, and how is the quality and quantity of the data?
- What might be the most significant impacts of a data science project?
- Is there any positive or negative impact on other teams?
- What computation resources are available for model training and model execution?
- Can we define key metrics to compare and quantify business value?
- Are there any data security, privacy, and legal concerns?
- What are the desired milestones, checkpoints, and timeline?
- Is the final application online or offline?
- Are the data sources online or offline?

It is likely to have a series of intense meetings and heated discussions to frame the project reasonably. After the planning stage, we should be able to define a set of key metrics related to the project, identify some offline and online data sources, request needed computation resources, draft a tentative timeline and milestones, and form a team of data scientist, data engineer, software developer, project manager and members from the business operation. Data scientists should play a significant role in these discussions. If data scientists do not lead the project formulation and planning, the project may not catch the desired timeline and milestones.

2.4.3 Project Modeling Stage

Even though we already set some strategies, milestones, and timelines at the problem formulation and project planning stage, data science projects are dynamic. There could be uncertainties along the road. As a data scientist, communicating any newly encountered difficulties or opportunities during the modeling stage to the entire team is essential to keep the data science project progress. Data cleaning, data wrangling, and exploratory data analysis are great starting points toward modeling with the available data source identified at the planning stage. Meanwhile, abstracting the business problem to be a set of statistical and machine learning problems is an iterative process. Business problems can rarely be solved by using just one statistical or machine learning model. Using a sequence of methods to decompose the business problem is one of the critical responsibilities for a senior data scientist. The process requires iterative rounds of discussions with the business and data engineering team based on each iteration's new learnings. Each iteration includes both data-related and model-related parts.

2.4.3.1 Data Related Part

Data cleaning, data preprocessing, and feature engineering are related procedures that aim to create usable variables or features for statistical and machine learning models. A critical aspect of data related procedures is to make sure the data source we are using is a good representation of the situation where the final trained model will be applied. The same representation is rarely possible, and it is ok to start with a reasonable approximation. A data scientist must be clear on the assumptions and communicate the limitations of biased data with the team and quantify its impact on the application. In the data-related part, sometimes the available data is not relevant to the business problem we want to solve. We have to collect more and relevant data before modeling.

2.4.3.2 Model Related Part

There are different types of statistical and machine learning models, such as supervised learning, unsupervised learning, and causal inference. For each type, there are various algorithms, libraries, or packages readily available. To solve a business problem, we sometimes need to piece together a few methods at the model exploring and developing stage. This stage also includes model training, validation, and testing to ensure the model works well in the production environment (i.e., it can be generalized well and not causing overfitting). The model selection follows Occam's razor, choosing the simplest among a set of compatible models. Before we try complicated models, it is good to get some benchmarks by additional business rules, common-sense decisions, or standard models (such as random forest for classification and regression problems).

2.4.4 Model Implementation and Post Production Stage

For offline application data science projects, the end product is often a detailed report with model results and output. However, for online application projects, a trained model is just halfway from the finish line. The offline data is stored and processed in a different environment from the online production environment. Building the online data pipeline and implementing machine learning models in a production environment requires lots of additional work. Even though recent advance in cloud infrastructure lowers the barrier dramatically, it still takes effort to implement an offline model in the online production system. Before we promote the model to production, there are two more steps to go:

1. Shadow mode
2. A/B testing

A **shadow mode** is like an observation period when the data pipeline and machine learning models run as fully functional, but we only record the model output without any actions. Some people call it proof of concept (POC). During the shadow mode, people frequently check the data pipeline and model and detect bugs such as a timeout, missing features, version conflict (for example, Python 2 vs. Python 3), data type mismatch, etc.

Once the online model passes the shadow mode, A/B testing is the next stage. During A/B testing, all the incoming observations are randomly separated into two groups: control and treatment. The control group will skip the machine learning model, while the treatment group is going through the machine learning model. After that, people monitor a list of pre-defined key metrics during a specific period to compare the control and treatment groups. The differences in these metrics determine whether the machine learning model provides business value or not. Real applications can be complicated. For example, there can be multiple treatment groups, or hundreds, even thousands of A/B testing running by different teams at any given time in the same production environment.

Once the A/B testing shows that the model provides significant business value, we can put it into full production. It is ideal that the model runs as expected and continues to offer scalable values. However, the business can change, and a machine learning model that works now can break tomorrow, and features available now may not be available tomorrow. We need a monitoring system to notify us when one or multiple features change. When the model performance degrades below a pre-defined level, we need to fine-tune the parameters and thresholds, re-train the model with more recent data, add or remove features to improve model performance. Eventually, any model will fail or retire at some time with a pre-defined model retirement plan.

2.4.5 Project Cycle Summary

Data science end-to-end project cycle is a complicated process that requires close collaboration among many teams. The data scientist, maybe the only scientist in the team, has to lead the planning discussion and model development based on data available and communicate key assumptions and uncertainties. A data science project may fail at any stage, and a clear end-to-end cycle view of the project helps avoid some mistakes.

2.5 Common Mistakes in Data Science

Data science projects can go wrong at different stages in many ways. Most textbooks and online blogs focus on technical mistakes about machine learning models, algorithms, or theories, such as detecting outliers and overfitting. It is important to avoid these technical mistakes. However, there are common systematic mistakes across data science projects that are rarely discussed in textbooks. In this section, we describe these common mistakes in detail so that readers can proactively identify and avoid these systematic mistakes in their data science projects.

2.5.1 Problem Formulation Stage

The most challenging part of a data science project is problem formulation. Data science project stems from pain points of the business. The draft version of the project's goal is relatively vague without much quantification or is the gut feeling of the leadership team. Often there are multiple teams involved in the initial project formulation stage, and they have different views. It is easy to have misalignment across teams, such as resource allocation, milestone deliverable, and timeline. Data science team members with technical backgrounds sometimes are not even invited to the initial discussion at the problem formulation stage. It sounds ridiculous, but sadly true that a lot of resources are spent on **solving the wrong problem**, the number one systematic common mistake in data science. Formulating a business problem into the right data science project requires an in-depth understanding of the business context, data availability and quality, computation infrastructure, and methodology to leverage the data to quantify business value.

We see people **over-promise about business value** all the time, another common mistake that will fail the project at the beginning. With big data and machine learning hype, leaders across many industries often have unrealistic high expectations on data science. It is especially true during enterprise transformation when there is a strong push to adopt new technology to get value out of the data. The unrealistic expectations are based on assumptions that are way off the chart without checking the data availability, data quality, computation resource, and current best practices in the field. Even when there is some exploratory analysis by the data science team at the problem formulation stage, project leaders sometimes ignore their data-driven voice.

These two systematic mistakes undermine the organization's data science strategy. The higher the expectation, the bigger the disappointment when the project cannot deliver business value. Data and business context are essential to formulate the business problem and set reachable business value. It helps avoid mistakes by having a strong data science leader with a broad technical background and letting data scientists coordinate and drive the problem formulation and set realistic goals based on data and business context.

2.5.2   Project Planning Stage

Now suppose the data science project is formulated correctly with a reasonable expectation on the business value. The next step is to plan the project by allocating resources, setting up milestones and timelines, and defining deliverables. In most cases, project managers coordinate different teams involved in the project and use agile project management tools similar to those in software development. Unfortunately, the project management team may not have experience with data science projects and hence fail to account for the uncertainties at the planning stage. The fundamental difference between data science projects and other projects leads to another common mistake: **too optimistic about the timeline**. For example, data exploratory and data preparation may take 60% to 80% of the total time for a given data science project, but people often don't realize that.

When there are a lot of data already collected across the organization, people assume we have enough data for everything. It leads to the mistake: **too optimistic about data availability and quality**. We need not "big data," but data that can help us solve the problem. The data available may be of low quality, and we need to put substantial effort into cleaning the data before we can use it. There are "unexpected" efforts to bring the right and relevant data for a specific data science project. To ensure smooth delivery of data science projects, we need to account for the "unexpected" work at the planning stage. Data scientists all know data preprocessing and feature engineering is usually the most time-consuming part of a data science project. However, people outside data science are not aware of it, and we need to educate other team members and the leadership team.

2.5.3 Project Modeling Stage

Finally, we start to look at the data and fit some models. One common mistake at this stage is **unrepresentative data**. The model trained using historical data may not generalize to the future. There is always a problem with biased or unrepresentative data. As a data scientist, we need to use data that are closer to the situation where the model will apply and quantify the impact of model output in production. Another mistake at this stage is **overfitting and obsession with complicated models**. Now, we can easily get hundreds or even thousands of features, and the machine learning models are getting more complicated. People can use open source libraries to try all kinds of models and are sometimes obsessed with complicated models instead of using the simplest among a set of compatible models with similar results.

The data used to build the models is always somewhat biased or unrepresentative. Simpler models are better to generalize. It has a higher chance of providing consistent business value once the model passes the test and is finally implemented in the production environment. The existing data and methods at hand may be insufficient to solve the business problem. In that case, we can try to collect more data, do feature engineering, or develop new models. However, if there is a fundamental gap between data and the business problem, the data scientist must make the tough decision to unplug the project.

On the other hand, data science projects usually have high visibility and may be initiated by senior leadership. Even after the data science team provided enough evidence that they can't deliver the expected business value, people may not want to stop the project, which leads to another common mistake at the modeling stage: **take too long to fail**. The earlier we can prevent a failing project, the better because we can put valuable resources into other promising projects. It damages the data science strategy, and everyone will be hurt by a long data science project that is doomed to fail.

2.5.4   Model Implementation and Post Production Stage

Now suppose we have found a model that works great for the training and testing data. If it is an online application, we are halfway. The next is to implement the model, which sounds like alien work for a data scientist without software engineering experience in the production system. The data engineering team can help with model production. However, as a data scientist, we need to know the potential mistakes at this stage. One big mistake is **missing shadow mode and A/B testing** and assuming that the model performance at model training/testing stays the same in the production environment. Unfortunately, the model trained and evaluated using historical data nearly never performs the same in the production environment. The data used in the offline training may be significantly different from online data, and the business context may have changed. If possible, machine learning models in production should always go through shadow mode and A/B testing to evaluate performance.

In the model training stage, people usually focus on model performance, such as accuracy, without paying too much attention to the model execution time. When a model runs online in real-time, each instance's total run time (i.e., model latency) should not impact the customer's user experience. Nobody wants to wait for even one second to see the results after click the "search" button. In the production stage, feature availability is crucial to run a real-time model. Engineering resources are essential for model production. However, in traditional companies, it is common that a data science project **fails to scale in real-time applications** due to lack of computation capacity, engineering resources, or non-tech culture and environment.

As the business problem evolves rapidly, the data and model in the production environment need to change accordingly, or the model's performance deteriorates over time. The online production environment is more complicated than model training and testing. For example, when we pull online features from different resources, some may be missing at a specific time; the model may run into a time-out zone, and various software can cause the version problem. We need regular checkups during the entire life of the model cycle from implementation to retirement. Unfortunately, people often don't set the monitoring system for data science projects, and it is another common mistake: **missing necessary online checkup**. It is essential to set a monitoring dashboard and automatic alarms, create model tuning, re-training, and retirement plans.

2.5.5 Summary of Common Mistakes

The data science project is a combination of art, science, and engineering. A data science project may fail in different ways. However, the data science project can provide significant business value if we put data and business context at the center of the project, get familiar with the data science project cycle and proactively identify and avoid these potential mistakes. Here is the summary of the mistakes:

- Solving the wrong problem
- Overpromise on business value
- Too optimistic about the timeline
- Too optimistic about data availability and quality
- Unrepresentative data
- Overfitting and obsession with complicated models
- Take too long to fail
- Missing A/B testing
- Fail to scale in real-time applications
- Missing necessary online checkup

3

Introduction to the Data

Before tackling analytics problem, we start by introducing data to be analyzed in later chapters.

3.1 Customer Data for a Clothing Company

Our first data set represents customers of a clothing company who sells products in physical stores and online. This data is typical of what one might get from a company's marketing data base (the data base will have more data than the one we show here). This data includes 1000 customers:

1. Demography

   - age: age of the respondent
   - gender: male/female
   - house: 0/1 variable indicating if the customer owns a house or not

2. Sales in the past year

   - store_exp: expense in store
   - online_exp: expense online
   - store_trans: times of store purchase
   - online_trans: times of online purchase

3. Survey on product preference

   It is common for companies to survey their customers and draw insights to guide future marketing activities. The survey is as below:

   How strongly do you agree or disagree with the following statements:

DOI: 10.1201/9781351132916-3   47

1. Strong disagree
2. Disagree
3. Neither agree nor disagree
4. Agree
5. Strongly agree

- Q1. I like to buy clothes from different brands
- Q2. I buy almost all my clothes from some of my favorite brands
- Q3. I like to buy premium brands
- Q4. Quality is the most important factor in my purchasing decision
- Q5. Style is the most important factor in my purchasing decision
- Q6. I prefer to buy clothes in store
- Q7. I prefer to buy clothes online
- Q8. Price is important
- Q9. I like to try different styles
- Q10. I like to make decision myself and don't need too much of others' suggestions

There are 4 segments of customers:

1. Price
2. Conspicuous
3. Quality
4. Style

Let's check it:

```
str(sim.dat,vec.len=3)
```

```
## 'data.frame':    1000 obs. of  19 variables:
##  $ age        : int  57 63 59 60 51 59 57 57 ...
##  $ gender     : chr  "Female" "Female" "Male" ...
##  $ income     : num  120963 122008 114202 113616 ...
##  $ house      : chr  "Yes" "Yes" "Yes" ...
```

```
## $ store_exp    : num   529 478 491 348 ...
## $ online_exp   : num   304 110 279 142 ...
## $ store_trans  : int   2 4 7 10 4 4 5 11 ...
## $ online_trans : int   2 2 2 2 4 5 3 5 ...
## $ Q1           : int   4 4 5 5 4 4 4 5 ...
## $ Q2           : int   2 1 2 2 1 2 1 2 ...
## $ Q3           : int   1 1 1 1 1 1 1 1 ...
## $ Q4           : int   2 2 2 3 3 2 2 3 ...
## $ Q5           : int   1 1 1 1 1 1 1 1 ...
## $ Q6           : int   4 4 4 4 4 4 4 4 ...
## $ Q7           : int   1 1 1 1 1 1 1 1 ...
## $ Q8           : int   4 4 4 4 4 4 4 4 ...
## $ Q9           : int   2 1 1 2 2 1 1 2 ...
## $ Q10          : int   4 4 4 4 4 4 4 4 ...
## $ segment      : chr   "Price" "Price" "Price" ...
```

Refer to Appendix for the simulation code.

3.2 Swine Disease Breakout Data

The swine disease data includes 120 simulated survey questions from 800 farms. There are three choices for each question. The outbreak status for the ith farm is generated from a $Bernoulli(1, p_i)$ distribution with p_i being a function of the question answers:

$$ln(\frac{p_i}{1 - p_i}) = \beta_0 + \Sigma_{g=1}^G \mathbf{x}_{i,g}^T \boldsymbol{\beta}_g$$

where β_0 is the intercept, $\mathbf{x}_{i,g}$ is a three-dimensional indication vector for question answer and $\boldsymbol{\beta}_g$ is the parameter vector corresponding to the gth predictor. Three types of questions are considered regarding their effects on the outcome. The first forty survey questions are important questions such that the coefficients of the three answers to these questions are all different:

$$\boldsymbol{\beta}_g = (1, 0, -1) \times \gamma, \ g = 1, ..., 40$$

The second forty survey questions are also important questions but only one answer has a coefficient that is different from the other two answers:

$$\beta_{\mathbf{g}} = (1, 0, 0) \times \gamma, \ g = 41, \dots, 80$$

The last forty survey questions are also unimportant questions such that all three answers have the same coefficients:

$$\beta_{\mathbf{g}} = (0, 0, 0) \times \gamma, \ g = 81, \dots, 120$$

The baseline coefficient β_0 is set to be $-\frac{40}{3}\gamma$ so that on average a farm have 50% of chance to have an outbreak. The parameter γ in the above simulation is set to control the strength of the questions' effect on the outcome. In this simulation study, we consider the situations where $\gamma = 0.1, 0.25, 0.5, 1, 2$. So the parameter settings are

$$\beta^{\mathbf{T}} = \left(\frac{40}{3}, \underbrace{1, 0, -1}_{question\ 1}, \dots, \underbrace{1, 0, 0}_{question\ 41}, \dots, \underbrace{0, 0, 0}_{question\ 81}, \dots, \underbrace{0, 0, 0}_{question\ 120} \right) * \gamma$$

For each value of γ, 20 data sets are simulated. The bigger γ is, the larger the corresponding parameter. We provided the data sets with $\gamma = 2$. Let's check the data:

```
disease_dat <- read.csv("http://bit.ly/2KXb1Qi")
# only show the last 7 columns here
head(subset(disease_dat,select=c("Q118.A","Q118.B","Q119.A",
                                 "Q119.B","Q120.A","Q120.B","y")))
```

```
##   Q118.A Q118.B Q119.A Q119.B Q120.A Q120.B y
## 1      1      0      0      0      0      1 1
## 2      0      1      0      1      0      0 1
## 3      1      0      0      0      1      0 1
## 4      1      0      0      0      0      1 1
## 5      1      0      0      0      1      0 0
## 6      1      0      0      1      1      0 1
```

Here y indicates the outbreak situation of the farms. y=1 means there is an outbreak in 5 years after the survey. The rest columns indicate survey responses. For example Q120.A = 1 means the respondent chose A in Q120. We consider C as the baseline.

Refer to Appendix for the simulation code.

3.3 MNIST Dataset

The MNIST dataset is a popular dataset for image classification machine learning model tutorials. It is conveniently included in the Keras library and ready to be loaded with build-in functions for analysis. The WIKI page of MNIST provides a detailed description of the dataset: https://en.wikipedia.org/wiki/MNIST_database. It contains 70,000 images of handwritten digits from American Census Bureau employees and American high school students. There are 60,000 training images and 10,000 testing images. Each image has a resolution of 28×28, and the numerical pixel values are in greyscale. Each image is represented by a 28×28 matrix with each element of the matrix an integer between 0 and 255. The label of each image is the intended digit of the handwritten image between 0 and 9. We cover the detailed steps to explore the MNIST dataset in the R and Python notebooks. A sample of the dataset is illustrated in figure 3.1:[1]

3.4 IMDB Dataset

The IMDB dataset (http://ai.stanford.edu/~amaas/data/sentiment/) is a popular dataset for text and language-related machine learning tutorials. It is also conveniently included in the Keras library, and there are a few build-in functions in Keras for data loading and pre-processing. It contains 50,000 movie reviews (25,000 in training

---

[1]The image is from https://en.wikipedia.org/wiki/File:MnistExamples.png

FIGURE 3.1
Sample of MNIST dataset

and 25,000 in testing) from IMDB, as well as each movie review's
binary sentiment: positive or negative. The raw data contains the
text of each movie review, and it has to be pre-processed before
being fitted with any machine learning models. By using Keras's
built-in functions, we can easily get the processed dataset (i.e.,
a numerical data frame) for machine learning algorithms. Keras'
build-in functions perform the following tasks to convert the raw
review text into a data frame:

1. Convert text data into numerical data. Machine learning
   models cannot work with raw text data directly, and we
   have to convert text into numbers. There are many different ways for the conversion and Keras' build-in function
   uses each word's rank of frequency in the entire training
   dataset to replace the raw text in both the training and
   testing dataset. For example, the 10th most frequent word
   is replaced by integer 10. There are a few additional setups
   for this process, including:

   a. Skip top frequent words. We usually skip a few top
      frequent words as they are mainly stopwords like "the"
      "and," or "a," which usually do not provide much
      information. There is a parameter in the build-in
      function to specify how many top words to skip.

b. Set the maximum number of unique words. The entire vocabulary of the unique words in the training dataset may be large, and many of them have very low frequencies such as just appearing once in the entire training dataset. To keep the size of the vocabulary, we can also set up the maximum number of the unique words using Keras' built-in function such that any words with least frequencies will be replaced with a special index such as "2".

2. Padding or truncation to keep all the reviews to be the same length. For most machine learning models, the algorithms expect to see the same number of features (i.e., same number of input columns in the data frame). There is a parameter in the Keras build-in function to set the maximum number of words in each review (i.e., `max_length`). For reviews that have less than `max_length` words, we pad them with "0". For reviews that have more than `max_length` words, we truncate them.

After the above pre-processing, each review is represented by one row in the data frame. There is one column for the binary positive/negative sentiment, and `max_length` columns input features converted from the raw review text. In the corresponding R and Python notebooks, we will go over the details of the data pre-processing using Keras' built-in functions.

4

Big Data Cloud Platform

Data has been a friend to statisticians and analysts for hundreds of years. Tabulated data is the most common format used daily. People used to store data on papers, tapes, diskettes, or hard drives. However, with the development of computer hardware and software, the volume, variety, and speed of the data have exceeded the capacity of traditional statisticians and analysts.

Therefore, using data has become a science that focuses on the question: how can we store, access, process, analyze the massive amount of data and provide actionable insights?

In the past few years, people have used commodity hardware and open-source software to create a big data ecosystem for data storage, data retrieval, and parallel computation. Hadoop and Spark have become popular platforms enabling data scientists, statisticians, and analysts to access the data and build models.

Programming skills in the big data platform have been an obstacle for traditional statisticians and analysts to become successful data scientists. However, cloud computing has reduced the difficulty significantly. The user interface of the data platform is much more user-friendly today, and many of the technical details are pushed to the background. Cloud systems also enable quick implementation of the production environment. Now data science emphasizes more on the data itself, models and algorithms on top of the data, rather than the platform, infrastructure and low-level programming such as Java.

DOI: 10.1201/9781351132916-4

4.1   Power of Cluster of Computers

We are familiar with our laptop/desktop computers which have three main components to do data computation: (1) Hard disk, (2) Memory, and (3) CPU.

The data and codes stored in the hard disk have specific features such as slow to read and write, and large capacity of around a few TB in today's market. Memory is fast to read and write but with small capacity in the order of a few dozens of GB in today's market. CPU is where all the computation happens.

For statistical software such as R, the amount of data it can process is limited by the computer's memory. The memory of computers before 2000 is less than 1 GB. The memory capacity grows way slower than the amount of the data. Now it is common that we need to analyze data far beyond the capacity of a single computer's memory, especially in an enterprise environment. Meanwhile, as the data size increases, to solve the same problem (such as regressions), the computation time is growing faster than linear. Using a cluster of computers become a common way to solve a big data problem. In figure 4.1 (right), a cluster of computers can be viewed as one powerful machine with memory, hard disk and CPU equivalent to

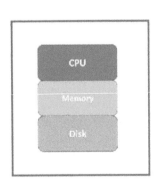
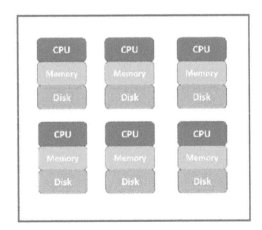

FIGURE 4.1
Single computer (left) and a cluster of computers (right)

the sum of individual computers. It is common to have hundreds or even thousands of nodes for a cluster.

In the past, users need to write code (such as MPI) to distribute data and do parallel computing. Fortunately, with the recent new development, the cloud environment for big data analysis is more user-friendly. As data is often beyond the size of the hard disk, the dataset itself is stored across different nodes (i.e., the Hadoop system). When doing analysis, the data is distributed across different nodes, and algorithms are parallel to leverage corresponding nodes' CPUs to compute (i.e., the Spark system).

4.2 Evolution of Cluster Computing

Using computer clusters to solve general-purpose data and analytics problems needs a lot of effort if we have to specifically control every element and steps such as data storage, memory allocation, and parallel computation. Fortunately, high tech companies and open source communities have developed the entire ecosystem based on Hadoop and Spark. Users need only to know high-level scripting languages such as Python and R to leverage computer clusters' distributed storage, memory and parallel computation power.

4.2.1 Hadoop

The very first problem internet companies face is that a lot of data has been collected and how to better store these data for future analysis. Google developed its own file system to provide efficient, reliable access to data using large clusters of commodity hardware. The open-source version is known as Hadoop Distributed File System (HDFS). Both systems use Map-Reduce to allocate computation across computation nodes on top of the file system. Hadoop is written in Java and writing map-reduce job using Java is a direct way to interact with Hadoop which is not familiar to many in the data and analytics community. To help better use the Hadoop system, an SQL-like data warehouse system called

Hive, and a scripting language for analytics interface called Pig were introduced for people with analytics background to interact with Hadoop system. Within Hive, we can create user-defined functions through R or Python to leverage the distributed and parallel computing infrastructure. Map-reduce on top of HDFS is the main concept of the Hadoop ecosystem. Each map-reduce operation requires retrieving data from hard disk, then performing the computation, and storing the result onto the disk again. So, jobs on top of Hadoop require a lot of disk operation which may slow down the entire computation process.

4.2.2 Spark

Spark works on top of a distributed file system including HDFS with better data and analytics efficiency by leveraging in-memory operations. Spark is more tailored for data processing and analytics and the need to interact with Hadoop directly is greatly reduced. The spark system includes an SQL-like framework called Spark SQL and a parallel machine learning library called MLlib Fortunately for many in the analytics community, Spark also supports R and Python. We can interact with data stored in a distributed file system using parallel computing across nodes easily with R and Python through the Spark API and do not need to worry about lower-level details of distributed computing. We will introduce how to use an R notebook to drive Spark computations.

4.3   Introduction of Cloud Environment

Even though Spark provides a solution for big data analytics, the maintenance of the computing cluster and Spark system requires a dedicated team. Historically for each organization the IT departments own the hardware and the regular maintenance. It usually takes months for a new environment to be built and the cost is high. Luckily, the time to deployment and cost are dramatically down due to the cloud computation trend. Now we can create a

Spark computing cluster in the cloud in a few minutes with the desired configuration and the user only pay when the cluster is up. Cloud computing environments enable smaller organizations to adopt big data analytics.

There are many cloud computing environments such as Amazon's AWS, Google cloud and Microsoft Azure which provide a complete list of functions for heavy-duty enterprise applications. For example, Netflix runs its business entirely on AWS without owning any data centers. For beginners, however, Databricks provides an easy to use cloud system for learning purposes. Databricks is a company founded by the creators of Apache Spark and it provides a user-friendly web-based notebook environment that can create a Spark cluster on the fly to run R/Python/Scala/SQL scripts. We will use Databricks' free community edition to run demos in this book. Please note, to help readers to get familiar with the Databricks cloud system, the content of this section is partially adopted from the following web pages:

- `https://docs.databricks.com/sparkr/sparklyr.html`
- `http://spark.rstudio.com/index.html`

4.3.1 Open Account and Create a Cluster

Anyone can apply for a free Databrick account through `https://databricks.com/try-databricks` and please make sure to choose the "**COMMUNITY EDITION**" which does not require payment information and will always be free. Once the community edition account is open and activated. Users can create a cluster computation environment with Spark. The computing cluster associated with community edition account is relatively small, but it is good enough for running all the examples in this book. The main user interface to the computing environment is notebook: a collection of cells that contains formatted text or codes. When a new notebook is created, user will need to choose the default programming language type (i.e. Python, R, Scala, or SQL) and every cells in the notebook will assume the default programming language. However, user can easily override the default selection of programming language by

FIGURE 4.2
Example of R notebook

adding %sql, %python, %r or %scala at the first line of each cell to
indicate the programming language in that cell. Allowing running
different cells with different programming language in the same
notebook enable user to have the flexibility to choose the best tools
for each task. User can also define a cell to be markdown cell by
adding %md at the first line of the cell. A markdown cell does not
performance computation and it is just a cell to show formatted
text. Well separated cells with computation, graph and formatted
text enable user to create easy to maintain reproducible reports.
The link to a video showing how to open Databricks account, how
to create a cluster, and how to create notebooks is included in the
book's website.

4.3.2 R Notebook

For this book, we will use R notebook for examples and demos
and the corresponding Python notebook will be available online
too. For an R notebook, it contains multiple cells, and, by default,
the content within each cell are R scripts. Usually, each cell is a
well-managed segment of a few lines of codes that accomplish a
specific task. For example, figure 4.2 shows the default cell for an R
notebook. We can type in R scripts and comments same as we are
using R console. By default, only the result from the last line will
be shown following the cell. However, you can use print() function
to output results for any lines. If we move the mouse to the middle
of the lower edge of the cell below the results, a "+" symbol will

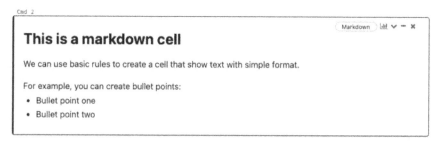

FIGURE 4.3
Example of Markdown cell

show up and click on the symbol will insert a new cell below. When we click any area within a cell, it will make it editable and you will see a few icons on the top right corn of the cell where we can run the cell, as well as add a cell below or above, copy the cell, cut the cell etc. One quick way to run the cell is Shift+Enter when the cell is chosen. User will become familiar with the notebook environment quickly.

4.3.3 Markdown cells

For an R notebook, every cell by default will contain R scripts. But if we put %md, %sql or %python at the first line of a cell, that cell becomes Markdown cell, SQL script cell, and Python script cell accordingly. For example, figure 4.3 shows a markdown cell with scripts and the actual appearance when exits editing mode. Markdown cell provides a straightforward way to descript what each cell is doing as well as what the entire notebook is about. It is a better way than a simple comment within the code.

4.4   Leverage Spark Using R Notebook

R is a powerful tool for data analysis given the data can be fit into memory. Because of the memory bounded dataset limit, R itself cannot be used directly for big data analysis where the data is likely stored in Hadoop and Spark system. By leverage the sparklyr package created by RStudio, we can use Databricks' R notebook to analyze data stored in the Spark system. As the data are stored across different nodes, Spark enables parallel computation using the collection of memory and CPU across all nodes. The fundamental data element in the Spark system is called Spark DataFrames (SDF). In this section, we will illustrate how to use Databricks' R notebook for big data analysis on top of the Spark environment through sparklyr package.

Install Package

First, we need to install sparklyr package which enables the connection between local node to Spark cluster environments. As it will install more than 10 dependencies, it may take a few minutes to finish. Be patient while it is installing! Once the installation finishes, load the sparklyr package as illustrated by the following code:

```
# Install sparklyr
if (!require("sparklyr")) {
install.packages("sparklyr")
}
# Load sparklyr package
library(sparklyr)
```

Create a Spark Connection

Once the library is loaded, we need to create a Spark Connection to link the computing node (i.e. local node) running the R notebook to the Spark environment. Here we use the "databricks" option for parameter method which is specific for databricks' cloud system. In other enterprise environments, please consult your administrator for details. The Spark Connection (i.e. sc) is the pipe to connect R notebook in the local node with the Spark Cluster. We can think of the R notebook is running on a local node that has its memory and CPU; the Spark system has a cluster of connected computation nodes, and the Spark Connection creates a mechanism to connect both systems. The Spark Connection can be established with:

```
# create a sparklyr connection
sc <- spark_connect(method = "databricks")
```

To simplify the learning process, let us use a very familiar small dataset: the iris dataset. It is part of the dplyr library and we can load that library to use the iris data frame. Now the iris dataset is still on the local node where the R notebook is running on. And we can check the first a few lines of the iris dataset using the code below:

```
library(dplyr)
```

```
##
## Attaching package: 'dplyr'

## The following objects are masked from 'package:stats':
##
##     filter, lag

## The following objects are masked from 'package:base':
##
##     intersect, setdiff, setequal, union
```

```
head(iris)
```

```
##    Sepal.Length Sepal.Width Petal.Length Petal.Width
## 1           5.1         3.5          1.4         0.2
## 2           4.9         3.0          1.4         0.2
## 3           4.7         3.2          1.3         0.2
## 4           4.6         3.1          1.5         0.2
## 5           5.0         3.6          1.4         0.2
## 6           5.4         3.9          1.7         0.4
##    Species
## 1  setosa
## 2  setosa
## 3  setosa
## 4  setosa
## 5  setosa
## 6  setosa
```

IMPORTANT - Copy Data to Spark Environment

In real applications, the data set may be massive and cannot fit in a single hard disk and most likely such data are already stored in the Spark system. If the data is already in Hadoop/Spark ecosystem in the form of SDF, we can create a local R object to link to the SDF by the tbl() function where my_sdf is the SDF in the Spark system, and my_sdf_tbl is the R local object that referring to my_sdf:

```
my_sdf_tbl <- tbl(sc, my_sdf)
```

As we just created a brand new Spark computing environment, there is no SDF in the system yet. We will need to copy a local dataset to the Spark environment. As we have already created the Spark Connection sc, it is easy to copy data to spark system using sdf_copy_to() function as below:

```
iris_tbl <- sdf_copy_to(sc = sc, x = iris, overwrite = T)
```

The above one-line code copies iris dataset from the local node to Spark cluster environment. "sc" is the Spark Connection we just created; "x" is the data frame that we want to copy; "overwrite" is the option whether we want to overwrite the target object if the same name SDF exists in the Spark environment. Finally, sdf_copy_to() function will return an R object representing the copied SDF (i.e. creating a "pointer" to the SDF such that we can refer iris_tbl in the R notebook to operate iris SDF). Now irir_tbl in the local R environment can be used to refer to the iris SDF in the Spark system.

To check whether the iris data was copied to the Spark environment successfully or not, we can use src_tbls() function to the Spark Connection (sc):

```
## code to return all the data frames associated with sc
src_tbls(sc)
```

Analyzing the Data

Now we have successfully copied the iris dataset to the Spark environment as a SDF. This means that iris_tbl is an R object representing the iris SDF and we can use iris_tbl in R to refer the iris dataset in the Spark system (i.e. the iris SDF). With the sparklyr packages, we can use nearly all the functions in dplyr to Spark DataFrame directly through iris_tbl, same as we are applying dplyr functions to a local R data frame in our laptop. For example, we can use the %>% operator to pass iris_tbl to the count() function:

```
iris_tbl %>% count
```

or using the head() function to return the first few rows in iris_tbl:

```
head(iris_tbl)
```

or applying more advanced data manipulation directly to iris_tbl:

```
iris_tbl %>%
    mutate(Sepal_Add = Sepal_Length + Sepal_Width) %>%
    group_by(Species) %>%
    summarize(count = n(), Sepal_Add_Avg = mean(Sepal_Add))
```

Collect Results Back to Local Node

Even though we can run nearly all of the dplyr functions on SDF, we cannot apply functions from other packages directly to SDF (such as ggplot()). For functions that can only work on local R data frames, we must copy the SDF back to the local node as an R data frame. To copy SDF back to the local node, we use the collect() function. The following code using collect() will collect the results of a few operations and assign the collected data to iris_summary, a local R data frame:

```
iris_summary <- iris_tbl %>%
    mutate(Sepal_Width_round = round(Sepal_Width * 2) / 2) %>%
    group_by(Species, Sepal_Width_round) %>%
    summarize(count = n(), Sepal_Length_avg = mean(Sepal_Length),
    Sepal_Length_stdev = sd(Sepal_Length)) %>%
    collect()
```

Now, iris_summary is a local R object to the R notebook and we can use any R packages and functions to it. In the following code, we will apply ggplot() to it, exactly the same as a stand along R console:

```
library(ggplot2)
ggplot(iris_summary, aes(Sepal_Width_round, Sepal_Length_avg,
        color = Species)) +
    geom_line(size = 1.2) +
    geom_errorbar(aes(ymin = Sepal_Length_avg - Sepal_Length_stdev,
                      ymax = Sepal_Length_avg + Sepal_Length_stdev),
                  width = 0.05) +
    geom_text(aes(label = count),
              vjust = -0.2,
              hjust = 1.2,
              color = "black") +
theme(legend.position="top")
```

In most cases, the heavy-duty data preprocessing and aggregation is done in Spark using functions in dplyr. Once the data is aggregated, the size is usually dramatically reduced and such reduced data can be collected to an R local object for downstream analysis.

Fit Regression to SDF

One of the advantages of the Spark system is the parallel machine learning algorithm. There are many statistical and machine learning algorithms developed to run in parallel across many CPUs with data across many memory units for SDF. In this example, we have already uploaded the iris data to the Spark system, and the data in the SDF can be referred through iris_tbl as in the last section. The linear regression algorithm implemented in the Spark system can be called through ml_linear_regression() function. The syntax to call the function is to define the local R object that representing the SDF (i.e. iris_tbl (local R object) for iris (SDF)), response variable (i.e. the y variable in linear regression in the SDF) and features (i.e. the x variables in linear regression in the SDF). Now, we can easily fit a linear regression for large dataset far beyond the memory limit of one single computer, and it is truly scalable and only constrained by the resource of the Spark cluster. Below is an illustration of how to fit a linear regression to SDF using R notebook:

```
fit1 <- ml_linear_regression(x = iris_tbl,
                response = "Sepal_Length",
                features = c("Sepal_Width", "Petal_Length",
                "Petal_Width"))
summary(fit1)
```

In the above code, x is the R object pointing to the SDF;
response is y-variable, features are the collection of explanatory
variables. For this function, both the data and computation are
in the Spark cluster which leverages multiple CPUs, distributed
memories and parallel computing.

Fit a K-means Cluster

Through the sparklyr package, we can use an R notebook to
access many Spark Machine Learning Library (MLlib) algorithms
such as Linear Regression, Logistic Regression, Survival Regression,
Generalized Linear Regression, Decision Trees, Random Forests,
Gradient-Boosted Trees, Principal Components Analysis, Naive-
Bayes, K-Means Clustering, and a few other methods. Below codes
fit a k-means cluster algorithm:

```
## Now fit a k-means clustering using iris_tbl data
## with only two out of four features in iris_tbl
fit2 <- ml_kmeans(x = iris_tbl, k = 3,
                    features = c("Petal_Length", "Petal_Width"))
# print our model fit
print(fit2)
```

After fitting the k-means model, we can apply the model to
predict other datasets through ml_predict() function. Following
code applies the model to iris_tbl again to predict the cluster
and collect the results as a local R object (i.e. prediction) using
collect() function:

```
prediction = collect(ml_predict(fit2, iris_tbl))
```

As `prediction` is a local R object, we can apply any R functions from any libraries to it. For example:

```
prediction %>%
  ggplot(aes(Petal_Length, Petal_Width)) +
  geom_point(aes(Petal_Width, Petal_Length,
              col = factor(prediction + 1)),
          size = 2, alpha = 0.5) +
  geom_point(data = fit2$centers, aes(Petal_Width, Petal_Length),
          col = scales::muted(c("red", "green", "blue")),
          pch = 'x', size = 12) +
  scale_color_discrete(name = "Predicted Cluster",
                  labels = paste("Cluster", 1:3)) +
labs(x = "Petal Length",
    y = "Petal Width",
    title = "K-Means Clustering",
    subtitle = "Use Spark ML to predict cluster
    membership with the iris dataset")
```

So far, we have illustrated

1. the relationship between a local node (i.e. where R notebook is running) and Spark Clusters (i..e where data are stored and computation are done);
2. how to copy a local data frame to a Spark DataFrames (please note if your data is already in Spark environment, there is no need to copy and we only need to build the connection. This is likely to be the case for enterprise environment);
3. how to manipulate Spark DataFrames for data cleaning and preprocessing through `dplyr` functions with the installation of `sparklyr` package;
4. how to fit statistical and machine learning models to Spark DataFrame in a truly parallel manner;
5. how to collect information from Spark DataFrames back to a local R object (i.e. local R data frame) for future analysis.

These procedures cover the basics of big data analysis that a data scientist needs to know as a beginner. We have an R notebook on the book website that contains the contents of this chapter. We also have a Python notebook on the book website.

4.5 Databases and SQL

4.5.1 History

Databases have been around for many years to efficiently organize, store, retrieve, and update data systematically. In the past, statisticians and analysts usually dealt with small datasets stored in text or spreadsheet files and often did not interact with database systems. Students from the traditional statistics department usually lack the necessary database knowledge. However, as data grow bigger, database knowledge becomes essential and required for statisticians, analysts and data scientists in an enterprise environment where data are stored in some form of database systems. Databases often contain a collection of tables and the relationship among these tables (i.e. schema). The table is the fundamental structure for databases that contain rows and columns similar to data frames in R or Python. Database management systems (DBMS) ensure data integration and security in real time operations. There are many different DBMS such as Oracle, SQL Server, MySQL, Teradata, Hive, Redshift and Hana. The majority of database operations are very similar among different DBMS, and Structured Query Language (SQL) is the standard language to use these systems.

SQL became a standard of the American National Standards Institute (ANSI) in 1986, and of the International Organization for Standardization (ISO) in 1987. The most recent version is published in December 2016. For typical users, fundamental knowledge is nearly the same across all database systems. In addition to the standard features, each DBMS providers include their own specific functions and features. So, for the same query, there may be slightly different implementations (i.e. SQL script) for different systems. In this section, we use the Databricks' SQL implementation (i.e. all the SQL scripts can run in Databricks SQL notebook).

More recent data is stored in a distributed system such as Hive for disk storage or Hana for in-memory storage. Most relational databases are row-based (i.e. data for each row are stored closely), whereas analytics workflows often favor column-based systems (i.e. data for each column are stored closely). Fortunately, as a database user, we only need to learn how to write SQL scripts to retrieve and manipulate data. Even though there are different implantations of SQL across different DBMS, SQL is nearly universal across relational databases including Hive and Spark, which means once we know SQL, our knowledge can be transferred among different database systems. SQL is easy to learn even if you do not have previous experience. In this session, we will go over the key concepts in the database and SQL.

4.5.2 Database, Table, and View

A database is a collection of tables that are related to each other. A database has its own database name and each table has its name as well. We can think a database is a "folder" where tables within a database are "files" within the folder. A table has rows and columns exactly as an R or Python pandas data frame. Each row (also called record) represents a unique instance of the subject and each column (also called field or attribute) represents a characteristic of the subject on the table. For each table, there is a special column called the primary key which uniquely identifies each of its records.

Tables within a specific database contain related information and the schema of a database illustrates all fields in every table as well as how these tables and fields relate to each other (i.e. the structure of a database). Tables can be filtered, joined and aggregated to return specific information. The view is a virtual table composed of fields from one or more base tables. The view does not store data and only store table structure. The view is also referred to as a saved query. The view is typically used to protect the data stored in the table and users can only query information from a view and cannot change or update its contents.

We will use two simple tables to illustrate basic SQL operations. These two tables are from an R dataset which contains the 50 states' population and income (`https://stat.ethz.ch/R-manual/R-patched/library/datasets/html/state.html`). The first table is called `divisions` which has two columns: `state` and `division` and the first few rows are shown in the following table:

state	division
Alabama	East South Central
Alaska	Pacific
Arizona	Mountain
Arkansas	West South Central
California	Pacific

The second table is called `metrics` which contains three columns: `state, population` and `income` and first few rows of the table are shown below:

state	population	income
Alabama	3615	3624
Alaska	365	6315
Arizona	2212	4530
Arkansas	2110	3378
California	21198	5114

To illustrate missing information, three more rows are added at the end of the original division table with state Alberta, Ontario, and Quebec with their corresponding division NULL. We first creat these two tables and save them as csv files, and then we upload these two files as Databricks tables.

4.5.3 Basic SQL Statement

After logging into Databricks and creating two tables, we can now create a notebook and choose the default language of the notebook to be SQL. There are a few very easy SQL statements to help us understand the database and table structure:

- `show database`: show current databases in the system
- `create database db_name`: create a new database with name `db_name`
- `drop database db_name`: delete database `db_name` (be careful when using it!)
- `use db_name`: set up the current database to be used
- `show tables`: show all the tables within the currently used database
- `describe tbl_name`: show the structure of table with name `tbl_name` (i.e. list of column name and data type)
- `drop tbl_name`: delete a table with name `tbl_name` (be careful when using it!)
- `select * from metrics limit 10`: show the first 10 rows of a table

If you are familiar with a procedural programming language such as C and FORTRAN or scripting languages such as R and Python, you may find SQL code a little bit strange. We should view SQL code by each specific chunk where it defines a specific task. SQL codes descript a specific task and DBMS will run and finish the task. SQL does not follow typical procedure program rules and we can think SQL is "descriptive" (i.e. we describe what we want using SQL and DBMS figures out how to do it).

4.5.3.1 SELECT Statement

SELECT is the most used statement in SQL, especially for database users and business analysts. It is used to extract specific information (i.e. column or columns) FROM one or multiple tables. It can be used to combine multiple tables. WHERE can be used in the SELECT statement to selected rows with specific conditions (i.e. filters). ORDER BY can be used in the SELECT statement to order the results in descending or ascending order of one or multiple columns. We can use * after SELECT to represent all columns in the table, or specifically write the column names separated by a comma. Below is the basic structure of a SELECT statement:

```
SELECT Col_Name1, Col_Name2
FROM Table_Name
```

```
WHERE Specific_Condition
ORDER BY Col_Name1, Col_Name2;
```

Here `Specific_Condition` is the typical logical conditions and only columns with `TRUE` for this condition will be chosen. For example, if we want to choose states and its total income where the population larger than 10000 and individual income less than 5000 with the result order by state name, we can use the following query:

```
SELECT state, income*    population AS total_income
FROM metrics
WHERE population > 10000 AND income < 5000
ORDER BY state
```

The `SELECT` statement is used to slicing and dicing the dataset as well as create new columns of interest (such as `total_income`) using basic computation functions.

4.5.3.2 Aggregation Functions and GROUP BY

We can also use aggregation functions in the `SELECT` statement to summarize the data. For example, `COUNT(col_name)` function will return the total number of not `NULL` rows for a specific column. Other aggregation function on numerical values include `MIN(col_name)`, `MAX(col_name)`, `AVG(col_name)`. Let's use the `metrics` table again to illustrate aggregation functions. For aggregation function, it takes all the rows that match WHERE condition (if any) and return one number. The following statement will calculate the maximum, minimum, and average population for all states starts with letter A to E.

```
SELECT sum(population) AS sum_pop, max(population) AS
max_pop, min(population) AS min_pop, avg(population)
AS avg_pop, count(population) AS count_pop
FROM metrics
WHERE substring(state, 1, 1) IN ('A', 'B', 'C', 'D', 'E')
```

The results from the above query only return one row as expected. Sometimes we want to find the aggregated value based on groups that can be defined by one or more columns. Instead of writing multiple SQL to calculate the aggregated value for each group, we can easily use the GROUP BY to calculate the aggregated value for each group in the SELECT statement. For example, if we want to find how many states in each division, we can use the following:

```
SELECT division, count(state) AS number_of_states
FROM divisions
GROUP BY division
```

Another special aggregation function is to return distinct values for one column or a combination of multiple columns. Simple use SELECT DISTINCT col_name1, col_name2 in the first line of the SELECT statement.

4.5.3.3 Join Multiple Tables

The database system is usually designed such that each table contains a piece of specific information and oftentimes we need to JOIN multiple tables to achieve a specific task. There are few types typically JOINs: inner join (keep only rows that match the join condition from both tables), left outer join (rows from inner join + unmatched rows from the first table), right outer join (rows from inner join + unmatched rows from the second table) and full outer join (rows from inner join + unmatched rows from both tables). The typical JOIN statement is illustrated below:

```
SELECT a.col_name1 AS var1, b.col_name2 AS var2
FROM tbl_one AS a
LEFT JOIN tabl_two AS b
ON a.col_to_match = b.col_to_match
```

For example, let us join the division table and metrics table to find what is the average population and income for each division, and the results order by division names:

```
SELECT a.division, avg(b.population) AS avg_pop,
avg(b.income) AS avg_inc
FROM divisions AS a
INNER JOIN metrics AS b
ON a.state = b.state
GROUP BY division
ORDER BY division
```

4.5.3.4   Add More Content into a Table

We can use the INSERT statement to add additional rows into a particular table, for example, we can add one more row to the metrics table by using the following query:

```
INSERT INTO metrics
VALUES ('Alberta', 4146, 7370)
```

4.5.4   Advanced Topics in Database

There are many advanced topics such as how to efficiently query data using index; how to take care of data integrity when multiple users are using the same table; algorithm behind data storage (i.e. column-wise or row-wise data storage); how to design the database schema. Users can learn these advanced topics gradually. We hope the basic knowledge covered in this section will kick off the initial momentum to learn SQL. As you can see, it is easy to write SQL statement to retrieve, join, slice, dice and aggregate data. The SQL notebook that contains all the above operations is included in the book's website.

5

Data Pre-processing

Many data analysis related books focus on models, algorithms and statistical inferences. However, in practice, raw data is usually not directly used for modeling. Data preprocessing is the process of converting raw data into clean data that is proper for modeling. A model fails for various reasons. One is that the modeler doesn't correctly preprocess data before modeling. Data preprocessing can significantly impact model results, such as imputing missing value and handling with outliers. So data preprocessing is a very critical part.

In real life, depending on the stage of data cleanup, data has the following types:

1. Raw data
2. Technically correct data
3. Data that is proper for the model
4. Summarized data
5. Data with fixed format

The raw data is the first-hand data that analysts pull from the database, market survey responds from your clients, the experimental results collected by the research and development department, and so on. These data may be very rough, and R sometimes can't read them directly. The table title could be multi-line, or the format does not meet the requirements:

- Use 50% to represent the percentage rather than 0.5, so R will read it as a character;
- The missing value of the sales is represented by "-" instead of space so that R will treat the variable as character or factor type;

DOI: 10.1201/9781351132916-5

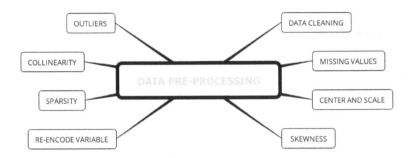

FIGURE 5.1
Data pre-processing outline

- The data is in a slideshow document, or the spreadsheet is not ".csv" but ".xlsx"
- ...

Most of the time, you need to clean the data so that R can import them. Some data format requires a specific package. Technically correct data is the data, after preliminary cleaning or format conversion, that R (or another tool you use) can successfully import it.

Assume we have loaded the data into R with reasonable column names, variable format and so on. That does not mean the data is entirely correct. There may be some observations that do not make sense, such as age is negative, the discount percentage is greater than 1, or data is missing. Depending on the situation, there may be a variety of problems with the data. It is necessary to clean the data before modeling. Moreover, different models have different requirements on the data. For example, some model may require the variables are of consistent scale; some may be susceptible to outliers or collinearity, some may not be able to handle categorical variables and so on. The modeler has to preprocess the data to make it proper for the specific model.

Sometimes we need to aggregate the data. For example, add up the daily sales to get annual sales of a product at different locations. In customer segmentation, it is common practice to build a profile for each segment. It requires calculating some statistics such as average age, average income, age standard deviation, etc.

Data aggregation is also necessary for presentation, or for data visualization.

The final table results for clients need to be in a nicer format than what used in the analysis. Usually, data analysts will take the results from data scientists and adjust the format, such as labels, cell color, highlight. It is important for a data scientist to make sure the results look consistent which makes the next step easier for data analysts.

It is highly recommended to store each step of the data and the R code, making the whole process as repeatable as possible. The R markdown reproducible report will be extremely helpful for that. If the data changes, it is easy to rerun the process. In the remainder of this chapter, we will show the most common data preprocessing methods.

Load the R packages first:

```
# install packages from CRAN
p_needed <- c('imputeMissings','caret','e1071','psych',
              'car','corrplot')
packages <- rownames(installed.packages())
p_to_install <- p_needed[!(p_needed %in% packages)]
if (length(p_to_install) > 0) {
    install.packages(p_to_install)
}

lapply(p_needed, require, character.only = TRUE)
```

5.1 Data Cleaning

After you load the data, the first thing is to check how many variables are there, the type of variables, the distributions, and data errors. Let's read and check the data:

```
sim.dat <- read.csv("http://bit.ly/2P5gTw4")
summary(sim.dat)
```

```
      age              gender           income          house          store_exp
 Min.   : 16.0    Female:554    Min.   : 41776    No :432    Min.   : -500
 1st Qu.: 25.0    Male  :446    1st Qu.: 85832    Yes:568    1st Qu.:  205
 Median : 36.0                  Median : 93869               Median :  329
 Mean   : 38.8                  Mean   :113543               Mean   : 1357
 3rd Qu.: 53.0                  3rd Qu.:124572               3rd Qu.:  597
 Max.   :300.0                  Max.   :319704               Max.   :50000
                                NA's   :184
   online_exp       store_trans      online_trans         Q1               Q2
 Min.   :  69     Min.   : 1.00    Min.   : 1.0     Min.   :1.0     Min.   :1.00
 1st Qu.: 420     1st Qu.: 3.00    1st Qu.: 6.0     1st Qu.:2.0     1st Qu.:1.00
 Median :1942     Median : 4.00    Median :14.0     Median :3.0     Median :1.00
 Mean   :2120     Mean   : 5.35    Mean   :13.6     Mean   :3.1     Mean   :1.82
 3rd Qu.:2441     3rd Qu.: 7.00    3rd Qu.:20.0     3rd Qu.:4.0     3rd Qu.:2.00
 Max.   :9479     Max.   :20.00    Max.   :36.0     Max.   :5.0     Max.   :5.00


       Q3               Q4               Q5               Q6               Q7
 Min.   :1.00     Min.   :1.00     Min.   :1.00     Min.   :1.00     Min.   :1.00
 1st Qu.:1.00     1st Qu.:2.00     1st Qu.:1.75     1st Qu.:1.00     1st Qu.:2.50
 Median :1.00     Median :3.00     Median :4.00     Median :2.00     Median :4.00
 Mean   :1.99     Mean   :2.76     Mean   :2.94     Mean   :2.45     Mean   :3.43
 3rd Qu.:3.00     3rd Qu.:4.00     3rd Qu.:4.00     3rd Qu.:4.00     3rd Qu.:4.00
 Max.   :5.00     Max.   :5.00     Max.   :5.00     Max.   :5.00     Max.   :5.00


       Q8               Q9              Q10               segment
 Min.   :1.0      Min.   :1.00     Min.   :1.00     Conspicuous:200
 1st Qu.:1.0      1st Qu.:2.00     1st Qu.:1.00     Price      :250
 Median :2.0      Median :4.00     Median :2.00     Quality    :200
 Mean   :2.4      Mean   :3.08     Mean   :2.32     Style      :350
 3rd Qu.:3.0      3rd Qu.:4.00     3rd Qu.:3.00
 Max.   :5.0      Max.   :5.00     Max.   :5.00
```

Are there any problems? Questionnaire response Q1-Q10 seem reasonable, the minimum is 1 and maximum is 5. Recall that the questionnaire score is 1-5. The number of store transactions (`store_trans`) and online transactions (`online_trans`) make sense too. Things to pay attention are

- There are some missing values.
- There are outliers for store expenses (`store_exp`). The maximum value is 50000. Who would spend $50000 a year buying clothes? Is it an imputation error?
- There is a negative value (-500) in `store_exp` which is not logical.
- Someone is 300 years old.

How to deal with that? Depending on the situation, if the sample size is large enough and the missing happens randomly, it does not hurt to delete those problematic samples. Or we can set these values as missing and impute them instead of deleting the rows.

```
# set problematic values as missings
sim.dat$age[which(sim.dat$age > 100)] <- NA
sim.dat$store_exp[which(sim.dat$store_exp < 0)] <- NA
# see the results
summary(subset(sim.dat, select = c("age", "store_exp")))
```

```
      age            store_exp
 Min.   :16.00   Min.   :  155.8
 1st Qu.:25.00   1st Qu.:  205.1
 Median :36.00   Median :  329.8
 Mean   :38.58   Mean   : 1358.7
 3rd Qu.:53.00   3rd Qu.:  597.4
 Max.   :69.00   Max.   :50000.0
 NA's   :1       NA's   :1
```

Now let's deal with the missing values in the data.

5.2 Missing Values

You can write a whole book about missing value. This section will only show some of the most commonly used methods without getting too deep into the topic. Chapter 7 of the book by De Waal, Pannekoek, and Scholtus (de Waal et al., 2011) makes a concise overview of some of the existing imputation methods. The choice of specific method depends on the actual situation. There is no best way.

One question to ask before imputation: Is there any auxiliary information? Being aware of any auxiliary information is critical. For example, if the system set customer who did not purchase as missing, then the real purchasing amount should be 0. Is missing a random occurrence? If so, it may be reasonable to impute with mean or median. If not, is there a potential mechanism for the missing data? For example, older people are more reluctant to disclose their ages in the questionnaire, so that the absence of age is not completely random. In this case, the missing values need to be estimated using the relationship between age and other independent variables. For example, use variables such as whether they have children, income, and other survey questions to build a model to predict age.

Also, the purpose of modeling is important for selecting imputation methods. If the goal is to interpret the parameter estimate or statistical inference, then it is important to study the missing mechanism carefully and to estimate the missing values using non-missing information as much as possible. If the goal is to predict, people usually will not study the absence mechanism rigorously (but sometimes the mechanism is obvious). If the absence mechanism is not clear, treat it as missing at random and use mean, median, or k-nearest neighbor to impute. Since statistical inference is sensitive to missing values, researchers from survey statistics have conducted in-depth studies of various imputation schemes which focus on valid statistical inference. The problem of missing values in the prediction model is different from that in the traditional survey. Therefore, there are not many papers on missing value imputation in the prediction model. Those who want to study further can refer to Saar-Tsechansky and Provost's comparison of different imputation methods (Maytal and Foster, 2007) and De Waal, Pannekoek and Scholtus' book (de Waal et al., 2011).

5.2.1 Impute Missing Values with Median/Mode

In the case of missing at random, a common method is to impute with the mean (continuous variable) or median (categorical variables). You can use `impute()` function in `imputeMissings` package.

```r
# save the result as another object
demo_imp <- impute(sim.dat, method = "median/mode")
# check the first 5 columns
# there is no missing values in other columns
summary(demo_imp[, 1:5])
```

age	gender	income	house	store_exp
Min. :16.00	Female:554	Min. : 41776	No :432	Min. : 155.8
1st Qu.:25.00	Male :446	1st Qu.: 87896	Yes:568	1st Qu.: 205.1
Median :36.00		Median : 93869		Median : 329.8
Mean :38.58		Mean :109923		Mean : 1357.7
3rd Qu.:53.00		3rd Qu.:119456		3rd Qu.: 597.3
Max. :69.00		Max. :319704		Max. :50000.0

After imputation, `demo_imp` has no missing value. This method is straightforward and widely used. The disadvantage is that it does not take into account the relationship between the variables. When there is a significant proportion of missing, it will distort the data. In this case, it is better to consider the relationship between variables and study the missing mechanism. In the example here, the missing variables are numeric. If the missing variable is a categorical/factor variable, the `impute()` function will impute with the mode.

You can also use `preProcess()` in package `caret`, but it is only for numeric variables, and cannot impute categorical variables. Since missing values here are numeric, we can use the `preProcess()` function. The result is the same as the `impute()` function. `PreProcess()` is a powerful function that can link to a variety of data preprocessing methods. We will use the function later for other data preprocessing.

```
imp <- preProcess(sim.dat, method = "medianImpute")
demo_imp2 <- predict(imp, sim.dat)
summary(demo_imp2[, 1:5])
```

age	gender	income	house	store_exp
Min. :16.00	Female:554	Min. : 41776	No :432	Min. : 155.8
1st Qu.:25.00	Male :446	1st Qu.: 87896	Yes:568	1st Qu.: 205.1
Median :36.00		Median : 93869		Median : 329.8
Mean :38.58		Mean :109923		Mean : 1357.7
3rd Qu.:53.00		3rd Qu.:119456		3rd Qu.: 597.3
Max. :69.00		Max. :319704		Max. : 50000.0

5.2.2 K-nearest Neighbors

K-nearest neighbor (KNN) will find the k closest samples (Euclidian distance) in the training set and impute the mean of those "neighbors."

Use `preProcess()` to conduct KNN:

```
imp <- preProcess(sim.dat, method = "knnImpute", k = 5)
# need to use predict() to get KNN result
demo_imp <- predict(imp, sim.dat)
# only show the first three elements
lapply(sim.dat, class)[1:3]
```

age	gender	income
Min. :-1.5910972	Female:554	Min. :-1.43989
1st Qu.:-0.9568733	Male :446	1st Qu.:-0.53732
Median :-0.1817107		Median :-0.37606
Mean : 0.0000156		Mean : 0.02389
3rd Qu.: 1.0162678		3rd Qu.: 0.21540
Max. : 2.1437770		Max. : 4.13627

The preProcess() in the first line will automatically ignore non-numeric columns.

Comparing the KNN result with the previous median imputation, the two are very different. This is because when you tell the preProcess() function to use KNN (the option method =" knnImpute"), it will automatically standardize the data. Another way is to use Bagging tree (in the next section). Note that KNN cannot impute samples with the entire row missing. The reason is straightforward. Since the algorithm uses the average of its neighbors if none of them has a value, what does it apply to calculate the mean?

Let's append a new row with all values missing to the original data frame to get a new object called temp. Then apply KNN to temp and see what happens:

```
temp <- rbind(sim.dat, rep(NA, ncol(sim.dat)))
imp <- preProcess(sim.dat, method = "knnImpute", k = 5)
demo_imp <- predict(imp, temp)
```

```
Error in FUN(newX[, i], ...) :
  cannot impute when all predictors are missing in the new data
  point
```

There is an error saying "cannot impute when all predictors are missing in the new data point". It is easy to fix by finding and removing the problematic row(s):

```
idx <- apply(temp, 1, function(x) sum(is.na(x)))
as.vector(which(idx == ncol(temp)))
```

It shows that row 1001 is problematic. You can go ahead to delete it.

5.2.3 Bagging Tree

Bagging (Bootstrap aggregating) was originally proposed by Leo Breiman. It is one of the earliest ensemble methods (Breiman, 1966). When used in missing value imputation, it will use the remaining variables as predictors to train a bagging tree and then use the tree to predict the missing values. Although theoretically, the method is powerful, the computation is much more intense than KNN. In practice, there is a trade-off between computation time and the effect. If a median or mean meet the modeling needs, even bagging tree may improve the accuracy a little, but the upgrade is so marginal that it does not deserve the extra time. The bagging tree itself is a model for regression and classification. Here we use prePprocess() to impute sim.dat:

```
imp <- preProcess(sim.dat, method = "bagImpute")
demo_imp <- predict(imp, sim.dat)
summary(demo_imp[, 1:5])
```

age	gender	income	house	store_exp
Min. :16.00	Female:554	Min. : 41776	No :432	Min. : 155.8
1st Qu.:25.00	Male :446	1st Qu.: 86762	Yes:568	1st Qu.: 205.1
Median :36.00		Median : 94739		Median : 329.0
Mean :38.58		Mean :114665		Mean : 1357.7
3rd Qu.:53.00		3rd Qu.:123726		3rd Qu.: 597.3
Max. :69.00		Max. :319704		Max. :50000.0

5.3   Centering and Scaling

It is the most straightforward data transformation. It centers and
scales a variable to mean 0 and standard deviation 1. It ensures
that the criterion for finding linear combinations of the predictors is based on how much variation they explain and therefore
improves the numerical stability. Models involving finding linear
combinations of the predictors to explain response/predictors variation need data centering and scaling, such as principle component
analysis (PCA) (Jolliffe, 2002), partial least squares (PLS) (Geladi
and Kowalski, 1986) and factor analysis (Mulaik, 2009). You can
quickly write code yourself to conduct this transformation.

Let's standardize the variable income from sim.dat:

```r
income <- sim.dat$income
# calculate the mean of income
mux <- mean(income, na.rm = T)
# calculate the standard deviation of income
sdx <- sd(income, na.rm = T)
# centering
tr1 <- income - mux
# scaling
tr2 <- tr1/sdx
```

Or the function preProcess() can apply this transformation to
a set of predictors.

```r
sdat <- subset(sim.dat, select = c("age", "income"))
# set the 'method' option
trans <- preProcess(sdat, method = c("center", "scale"))
# use predict() function to get the final result
transformed <- predict(trans, sdat)
```

Now the two variables are in the same scale. You can check
the result using summary(transformed). Note that there are missing
values.

5.4   Resolve Skewness

Skewness is defined to be the third standardized central moment. The formula for the sample skewness statistics is:

$$skewness = \frac{\sum(x_i - \bar{x})^3}{(n-1)v^{3/2}}$$

$$v = \frac{\sum(x_i - \bar{x})^2}{(n-1)}$$

A zero skewness means that the distribution is symmetric, i.e. the probability of falling on either side of the distribution's mean is equal.

There are different ways to remove skewness such as log, square root or inverse transformation. However, it is often difficult to determine from plots which transformation is most appropriate for correcting skewness. The Box-Cox procedure automatically identified a transformation from the family of power transformations that are indexed by a parameter λ (Box and Cox, 1964).

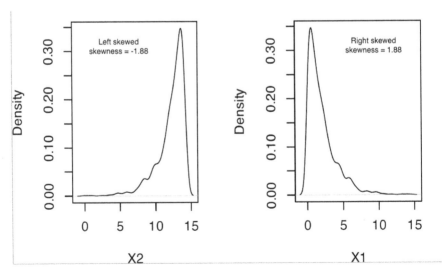

FIGURE 5.2
Example of skewed distributions

$$x^* = \begin{cases} \frac{x^\lambda - 1}{\lambda} & if \; \lambda \neq 0 \\ log(x) & if \; \lambda = 0 \end{cases}$$

It is easy to see that this family includes log transformation ($\lambda = 0$), square transformation ($\lambda = 2$), square root ($\lambda = 0.5$), inverse ($\lambda = -1$) and others in-between. We can still use function preProcess() in package caret to apply this transformation by chaning the method argument.

```
describe(sim.dat)
```

	vars	n	mean	sd	median	trimmed	mad	...
age	1	1000	38.84	16.42	36	37.69	16.31	
gender*	2	1000	1.45	0.50	1	1.43	0.00	
income	3	816	113543.07	49842.29	93869	104841.94	28989.47	
house*	4	1000	1.57	0.50	2	1.58	0.00	
store_exp	5	1000	1356.85	2774.40	329	839.92	196.45	
online_exp	6	1000	2120.18	1731.22	1942	1874.51	1015.21	
store_trans	7	1000	5.35	3.70	4	4.89	2.97	
online_trans	8	1000	13.55	7.96	14	13.42	10.38	

...

It is easy to see the skewed variables. If mean and trimmed differ a lot, there is very likely outliers. By default, trimmed reports mean by dropping the top and bottom 10%. It can be adjusted by setting argument trim=. It is clear that store_exp has outliers.

As an example, we will apply Box-Cox transformation on store_trans and online_trans:

```
# select the two columns and save them as dat_bc
dat_bc <- subset(sim.dat, select = c("store_trans", "online_trans"))
(trans <- preProcess(dat_bc, method = c("BoxCox")))
```

```
## Created from 1000 samples and 2 variables
##
## Pre-processing:
##    - Box-Cox transformation (2)
##    - ignored (0)
##
## Lambda estimates for Box-Cox transformation:
## 0.1, 0.7
```

The last line of the output shows the estimates of λ for each variable. As before, use `predict()` to get the transformed result:

```
transformed <- predict(trans, dat_bc)
par(mfrow = c(1, 2), oma = c(2, 2, 2, 2))
hist(dat_bc$store_trans, main = "Before Transformation",
    xlab = "store_trans")
hist(transformed$store_trans, main = "After Transformation",
    xlab = "store_trans")
```

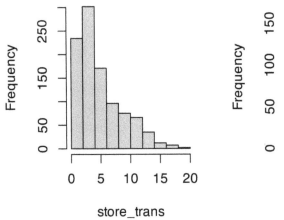 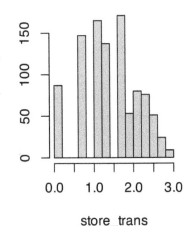

Before the transformation, the `stroe_trans` is skewed right. `BoxCoxTrans ()` can also conduct Box-Cox transform. But note that

`BoxCoxTrans ()` can only be applied to a single variable, and it is not possible to transform difference columns in a data frame at the same time.

```
(trans <- BoxCoxTrans(dat_bc$store_trans))
```

```
## Box-Cox Transformation
##
## 1000 data points used to estimate Lambda
##
## Input data summary:
##     Min. 1st Qu.  Median    Mean 3rd Qu.    Max.
##     1.00    3.00    4.00    5.35    7.00   20.00
##
## Largest/Smallest: 20
## Sample Skewness: 1.11
##
## Estimated Lambda: 0.1
## With fudge factor, Lambda = 0 will be used for transformations
```

```
transformed <- predict(trans, dat_bc$store_trans)
skewness(transformed)
```

```
## [1] -0.2155
```

The estimate of λ is the same as before (0.1). The skewness of the original observation is 1.1, and -0.2 after transformation. Although it is not strictly 0, it is greatly improved.

5.5 Resolve Outliers

Even under certain assumptions we can statistically define outliers, it can be hard to define in some situations. Box plot, histogram and some other basic visualizations can be used to initially check whether there are outliers. For example, we can visualize numerical non-survey variables in `sim.dat`:

```
# select numerical non-survey data
sdat <- subset(sim.dat, select = c("age", "income", "store_exp",
    "online_exp", "store_trans", "online_trans"))
# use scatterplotMatrix() function from car package
par(oma = c(2, 2, 1, 2))
car::scatterplotMatrix(sdat, diagonal = TRUE)
```

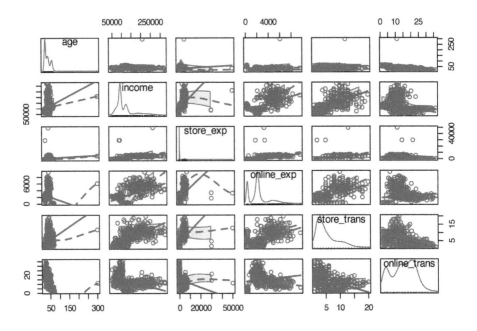

It is also easy to observe the pair relationship from the plot. `age` is negatively correlated with `online_trans` but positively correlated with `store_trans`. It seems that older people tend to purchase from the local store. The amount of expense is positively correlated with income. Scatterplot matrix like this can reveal lots of information before modeling.

In addition to visualization, there are some statistical methods to define outliers, such as the commonly used Z-score. The Z-score for variable Y is defined as:

$$Z_i = \frac{Y_i - \bar{Y}}{s}$$

where $\bar{Y}$ and s are mean and standard deviation for Y. Z-score is a measurement of the distance between each observation and the mean. This method may be misleading, especially when the sample size is small. Iglewicz and Hoaglin proposed to use the modified Z-score to determine the outlier (Iglewicz and Hoaglin, 1993):

$$M_i = \frac{0.6745(Y_i - \bar{Y})}{MAD}$$

Where MAD is the median of a series of $|Y_i - \bar{Y}|$, called the median of the absolute dispersion. Iglewicz and Hoaglin suggest that the points with the Z-score greater than 3.5 corrected above are possible outliers. Let's apply it to income:

```
# calculate median of the absolute dispersion for income
ymad <- mad(na.omit(sdat$income))
# calculate z-score
zs <- (sdat$income - mean(na.omit(sdat$income)))/ymad
# count the number of outliers
sum(na.omit(zs > 3.5))
```

```
## [1] 59
```

According to modified Z-score, variable income has 59 outliers. Refer to (Iglewicz and Hoaglin, 1993) for other ways of detecting outliers.

The impact of outliers depends on the model. Some models are sensitive to outliers, such as linear regression, logistic regression. Some are pretty robust to outliers, such as tree models, support vector machine. Also, the outlier is not wrong data. It is real

observation so cannot be deleted at will. If a model is sensitive
to outliers, we can use *spatial sign transformation* (Serneels et al.,
2006) to minimize the problem. It projects the original sample
points to the surface of a sphere by:

$$x_{ij}^* = \frac{x_{ij}}{\sqrt{\sum_{j=1}^{p} x_{ij}^2}}$$

where x_{ij} represents the ith observation and j^{th} variable. As shown
in the equation, every observation for sample i is divided by its
square mode. The denominator is the Euclidean distance to the
center of the p-dimensional predictor space. Three things to pay
attention here:

1. It is important to center and scale the predictor data
   before using this transformation
2. Unlike centering or scaling, this manipulation of the predictors transforms them as a group
3. If there are some variables to remove (for example, highly
   correlated variables), do it before the transformation

Function `spatialSign()` in `caret` package can conduct the transformation. Take `income` and `age` as an example:

```
# KNN imputation
sdat <- sim.dat[, c("income", "age")]
imp <- preProcess(sdat, method = c("knnImpute"), k = 5)
sdat <- predict(imp, sdat)
transformed <- spatialSign(sdat)
transformed <- as.data.frame(transformed)
par(mfrow = c(1, 2), oma = c(2, 2, 2, 2))
plot(income ~ age, data = sdat, col = "blue", main = "Before")
plot(income ~ age, data = transformed, col = "blue", main = "After")
```

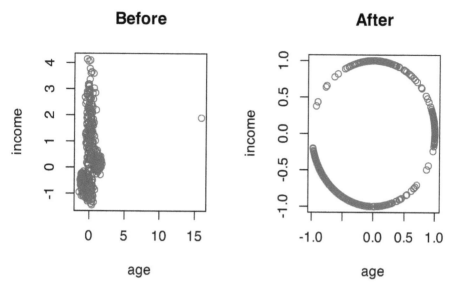

Some readers may have found that the above code does not seem to standardize the data before transformation. Recall the introduction of KNN, `prePorcess()` with `method="knnImpute"` by default will standardize data.

5.6 Collinearity

It is probably the technical term known by the most un-technical people. When two predictors are very strongly correlated, including both in a model may lead to confusion or problem with a singular matrix. There is an excellent function in `corrplot` package with the same name `corrplot()` that can visualize correlation structure of a set of predictors. The function has the option to reorder the variables in a way that reveals clusters of highly correlated ones.

```
# select non-survey numerical variables
sdat <- subset(sim.dat, select = c("age", "income", "store_exp",
    "online_exp", "store_trans", "online_trans"))
# use bagging imputation here
```

```
imp <- preProcess(sdat, method = "bagImpute")
sdat <- predict(imp, sdat)
# get the correlation matrix
correlation <- cor(sdat)
# plot
par(oma = c(2, 2, 2, 2))
corrplot.mixed(correlation, order = "hclust", tl.pos = "lt",
    upper = "ellipse")
```

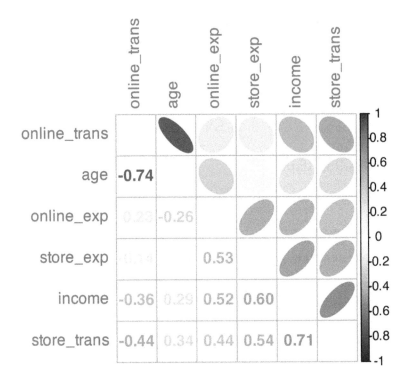

The closer the correlation is to 0, the lighter the color is and the closer the shape is to a circle. The elliptical means the correlation is not equal to 0 (because we set the upper = "ellipse"), the greater the correlation, the narrower the ellipse. Blue represents a positive correlation; red represents a negative correlation. The direction of the ellipse also changes with the correlation. The correlation coefficient is shown in the lower triangle of the matrix.

The variables relationship from previous scatter matrix are clear here: the negative correlation between age and online shopping, the positive correlation between income and amount of purchasing. Some correlation is very strong (such as the correlation between online_trans and age is -0.7414) which means the two variables contain duplicate information.

Section 3.5 of "Applied Predictive Modeling" (Kuhn and Johnston, 2013) presents a heuristic algorithm to remove a minimum number of predictors to ensure all pairwise correlations are below a certain threshold:

---

(1) Calculate the correlation matrix of the predictors.
(2) Determine the two predictors associated with the largest absolute pairwise correlation (call them predictors A and B).
(3) Determine the average correlation between A and the other variables. Do the same for predictor B.
(4) If A has a larger average correlation, remove it; otherwise, remove predictor B.
(5) Repeat Step 2-4 until no absolute correlations are above the threshold.

The findCorrelation() function in package caret will apply the above algorithm.

```
(highCorr <- findCorrelation(cor(sdat), cutoff = 0.7))
```

```
## [1] 2 6
```

It returns the index of columns need to be deleted. It tells us that we need to remove the 2nd and 6th columns to make sure the correlations are all below 0.7.

```
# delete highly correlated columns
sdat <- sdat[-highCorr]
# check the new correlation matrix
(cor(sdat))
```

The absolute value of the elements in the correlation matrix after removal are all below 0.7. How strong does a correlation have to get, before you should start worrying about multicollinearity? There is no easy answer to that question. You can treat the threshold as a tuning parameter and pick one that gives you best prediction accuracy.

5.7 Sparse Variables

Other than the highly related predictors, predictors with degenerate distributions can cause the problem too. Removing those variables can significantly improve some models' performance and stability (such as linear regression and logistic regression but the tree based model is impervious to this type of predictors). One extreme example is a variable with a single value which is called zero-variance variable. Variables with very low frequency of unique values are near-zero variance predictors. In general, detecting those variables follows two rules:

- The fraction of unique values over the sample size
- The ratio of the frequency of the most prevalent value to the frequency of the second most prevalent value.

nearZeroVar() function in the caret package can filter near-zero variance predictors according to the above rules. In order to show the useage of the function, let's arbitaryly add some problematic variables to the origional data sim.dat:

```
# make a copy
zero_demo <- sim.dat
# add two sparse variable zero1 only has one unique value zero2 is a
# vector with the first element 1 and the rest are 0s
zero_demo$zero1 <- rep(1, nrow(zero_demo))
zero_demo$zero2 <- c(1, rep(0, nrow(zero_demo) - 1))
```

The function will return a vector of integers indicating which columns to remove:

```
nearZeroVar(zero_demo,freqCut = 95/5, uniqueCut = 10)
```

```
## [1] 20 21
```

As expected, it returns the two columns we generated. You can go ahead to remove them. Note the two arguments in the function `freqCut =` and `uniqueCut =` are corresponding to the previous two rules.

- `freqCut`: the cutoff for the ratio of the most common value to the second most common value
- `uniqueCut`: the cutoff for the percentage of distinct values out of the number of total samples

5.8 Re-encode Dummy Variables

A dummy variable is a binary variable $(0/1)$ to represent subgroups of the sample. Sometimes we need to recode categories to smaller bits of information named "dummy variables." For example, some questionnaires have five options for each question, A, B, C, D, and E. After you get the data, you will usually convert the corresponding categorical variables for each question into five nominal variables, and then use one of the options as the baseline.

Let's encode `gender` and `house` from `sim.dat` to dummy variables. There are two ways to implement this. The first is to use `class.ind()` from `nnet` package. However, it only works on one variable at a time.

```
dumVar <- nnet::class.ind(sim.dat$gender)
head(dumVar)
```

```
##        Female Male
## [1,]        1    0
## [2,]        1    0
## [3,]        0    1
## [4,]        0    1
## [5,]        0    1
## [6,]        0    1
```

Since it is redundant to keep both, we need to remove one of them when modeling. Another more powerful function is `dummyVars()` from `caret`:

```
# use "origional variable name + level" as new name
dumMod <- dummyVars(~gender + house + income,
                data = sim.dat,
                levelsOnly = F)
head(predict(dumMod, sim.dat))
```

```
##   genderFemale genderMale houseNo houseYes income
## 1            1          0       0        1 120963
## 2            1          0       0        1 122008
## 3            0          1       0        1 114202
## 4            0          1       0        1 113616
## 5            0          1       0        1 124253
## 6            0          1       0        1 107661
```

dummyVars() can also use formula format. The variable on the right-hand side can be both categorical and numeric. For a numerical variable, the function will keep the variable unchanged. The advantage is that you can apply the function to a data frame without removing numerical variables. Other than that, the function can create interaction term:

```
dumMod <- dummyVars(~gender + house + income + income:gender,
               data = sim.dat,
               levelsOnly = F)
head(predict(dumMod, sim.dat))
```

```
##    genderFemale genderMale houseNo houseYes income
## 1             1          0       0        1 120963
## 2             1          0       0        1 122008
## 3             0          1       0        1 114202
## 4             0          1       0        1 113616
## 5             0          1       0        1 124253
## 6             0          1       0        1 107661
##    genderFemale:income genderMale:income
## 1               120963                 0
## 2               122008                 0
## 3                    0            114202
## 4                    0            113616
## 5                    0            124253
## 6                    0            107661
```

If you think the impact income levels on purchasing behavior is different for male and female, then you may add the interaction term between income and gender. You can do this by adding income: gender in the formula.

6

Data Wrangling

This chapter focuses on some of the most frequently used data manipulations and shows how to implement them in R and Python. It is critical to explore the data with descriptive statistics (mean, standard deviation, etc.) and data visualization before analysis. Transform data so that the data structure is in line with the requirements of the model. You also need to summarize the results after analysis.

When the data is too large to fit in a computer's memory, we can use some big data analytics engine like Spark on a cloud platform (see Chapter 4). Even the user interface of many data platforms is much more friendly now, it is still easier to manipulate the data as a local data frame. Spark's R and Python interfaces aim to keep the data manipulation syntax consistent with popular packages for local data frames. As shown in Section 4.4, we can run nearly all of the dplyr functions on a spark data frame once setting up the Spark environment. And the Python interface pyspark uses a similar syntax as pandas. This chapter focuses on data manipulations on standard data frames, which is also the foundation of big data manipulation.

Even when the data can fit in the memory, there may be a situation where it is slow to read and manipulate due to a relatively large size. Some R packages can make the process faster with the cost of familiarity, especially for data wrangling. But it avoids the hurdle of setting up Spark cluster and working in an unfamiliar environment. It is not a topic in this chapter but Appendix A briefly introduces some of the alternative R packages to read, write and wrangle a data set that is relatively large but not too big to fit in the memory.

There are many fundamental data processing functions in R. They lack consistent coding and can't flow together easily. Learning

DOI: 10.1201/9781351132916-6

all of them is a daunting task and unnecessary. R Studio developed a collection of packages and bundled them in `tidyverse` to systemize data wrangling and analysis tasks. You can see the package list in `tidyverse` on the website[1]. This chapter focuses on some of the `tidyverse` packages to do data wrangling for the following reasons:

- Those packages are widely used among R users in data science.
- The code is more efficient.
- The code syntax is consistent, which makes it easier to remember and read.

Section 6.1.2 introduces some base R functions outside the `tidyverse` universe, such as `apply()`, `lapply()` and `sapply()`. They are complementary functions when you are working with a data frame.

Load the R packages first:

```
# install packages from CRAN
p_needed <- c('dplyr','tidyr')
packages <- rownames(installed.packages())
p_to_install <- p_needed[!(p_needed %in% packages)]
if (length(p_to_install) > 0) {
    install.packages(p_to_install)
}

lapply(p_needed, require, character.only = TRUE)
```

6.1 Summarize Data

6.1.1 `dplyr` Package

`dplyr` provides a flexible grammar of data manipulation focusing on tools for working with data frames (hence the `d` in the name). It is faster and more friendly:

---

[1] https://www.tidyverse.org/packages/

- It identifies the most important data manipulations and make them easy to use from R.
- It performs faster for in-memory data by writing key pieces in C++ using `Rcpp`.
- The interface is the same for data frame, data table or database.

We will illustrate the following functions in order using the clothing company data:

1. Display
2. Subset
3. Summarize
4. Create new variable
5. Merge

```
# Read data
sim.dat <- read.csv("http://bit.ly/2P5gTw4")
```

6.1.1.1 Display

- `tbl_df()`: Convert the data to `tibble` which offers better checking and printing capabilities than traditional data frames. It will adjust output width according to fit the current window.

```
tbl_df(sim.dat)
```

- `glimpse()`: This is like a transposed version of `tbl_df()`

```
glimpse(sim.dat)
```

6.1.1.2 Subset

Get rows with `income` more than 300000:

```
filter(sim.dat, income >300000) %>%
  tbl_df()
```

Here we use the operator %>%. It is called a "pipe operator" which pipes a value forward into an expression or function call. What you get in the left operation will be the first argument or the only argument in the right operation.

```
x %>% f(y) = f(x, y)
y %>% f(x, ., z) = f(x, y, z )
```

It is an operator from magrittr which can be really beneficial. The following R code is difficulty to read and understand without using the pipe operator.

```
ave_exp <- filter(
  summarise(
    group_by(
      filter(
        sim.dat,
        !is.na(income)
      ),
      segment
    ),
    ave_online_exp = mean(online_exp),
    n = n()
  ),
  n > 200
)
```

The same function with pipe operator "%>%":

```
ave_exp <- sim.dat %>%
  filter(!is.na(income)) %>%
  group_by(segment) %>%
  summarise(
    ave_online_exp = mean(online_exp),
    n = n() ) %>%
  filter(n > 200)
```

It is much easier to read:

1. Delete observations from `sim.dat` with missing income values
2. Group the data from step 1 by variable `segment`
3. Calculate mean of online expense for each segment and save the result as a new variable named `ave_online_exp`
4. Calculate the size of each segment and saved it as a new variable named `n`
5. Get segments with size larger than 200

You can use `distinct()` to delete duplicated rows.

```
dplyr::distinct(sim.dat)
```

`sample_frac()` will randomly select some rows with a specified percentage. `sample_n()` can randomly select rows with a specified number.

```
dplyr::sample_frac(sim.dat, 0.5, replace = TRUE)
dplyr::sample_n(sim.dat, 10, replace = TRUE)
```

slice() will select rows by position:

```
dplyr::slice(sim.dat, 10:15)
```

It is equivalent to sim.dat[10:15,].
top_n() will select the order top n entries:

```
dplyr::top_n(sim.dat,2,income)
```

If you want to select columns instead of rows, you can use
select(). The following are some sample codes:

```
# select by column name
dplyr::select(sim.dat,income,age,store_exp)

# select columns whose name contains a character string
dplyr::select(sim.dat, contains("_"))

# select columns whose name ends with a character string
# similar there is "starts_with"
dplyr::select(sim.dat, ends_with("e"))

# select columns Q1,Q2,Q3,Q4 and Q5
select(sim.dat, num_range("Q", 1:5))

# select columns whose names are in a group of names
dplyr::select(sim.dat, one_of(c("age", "income")))

# select columns between age and online_exp
dplyr::select(sim.dat, age:online_exp)

# select all columns except for age
dplyr::select(sim.dat, -age)
```

6.1.1.3 Summarize

Let us use a standard marketing problem, customer segmentation, to illustrate how to summarize data. It usually starts with designing survey and collecting data. Then run a cluster analysis on the data to get customer segments. Once we have different segments, the next is to understand how each group of customer look like by summarizing some key metrics. For example, we can do the following data aggregation for different segments of clothes customers.

```r
dat_summary <- sim.dat %>%
  dplyr::group_by(segment) %>%
  dplyr::summarise(Age = round(mean(na.omit(age)), 0),
          FemalePct = round(mean(gender == "Female"), 2),
          HouseYes = round(mean(house == "Yes"), 2),
          store_exp = round(mean(na.omit(store_exp),
                              trim = 0.1), 0),
          online_exp = round(mean(online_exp), 0),
          store_trans = round(mean(store_trans), 1),
          online_trans = round(mean(online_trans), 1))

# transpose the data frame for showing purpose
# due to the limit of output width
cnames <- dat_summary$segment
dat_summary <- dplyr::select(dat_summary, - segment)
tdat_summary <- t(dat_summary) %>% data.frame()
names(tdat_summary) <- cnames
tdat_summary
```

```
##                Conspicuous  Price Quality   Style
## Age                  42.00  60.00   35.00   24.00
## FemalePct             0.32   0.45    0.47    0.81
## HouseYes              0.86   0.94    0.34    0.27
## store_exp          4990.00 501.00  301.00  200.00
## online_exp         4898.00 205.00 2013.00 1962.00
## store_trans          10.90   6.10    2.90    3.00
## online_trans         11.10   3.00   16.00   21.10
```

Now, let's look at the code in more details.

The first line `sim.dat` is easy. It is the data you want to work on. The second line `group_by(segment)` tells R that in the following steps you want to summarise by variable `segment`. Here we only summarize data by one categorical variable, but you can group by multiple variables, such as `group_by(segment, house)`. The third argument `summarise` tells R the manipulation(s) to do. Then list the exact actions inside `summarise()`. For example, `Age = round(mean(na.omit(age)),0)` tell R the following things:

1. Calculate the mean of column `age` ignoring missing value for each customer segment
2. Round the result to the specified number of decimal places
3. Store the result in a new variable named `Age`

The rest of the command above is similar. In the end, we calculate the following for each segment:

1. `Age`: average age for each segment
2. `FemalePct`: percentage for each segment
3. `HouseYes`: percentage of people who own a house
4. `stroe_exp`: average expense in store
5. `online_exp`: average expense online
6. `store_trans`: average times of transactions in the store
7. `online_trans`: average times of online transactions

There is a lot of information you can extract from those simple averages.

- Conspicuous: average age is about 40. It is a group of middle-age wealthy people. 1/3 of them are female, and 2/3 are male. They buy regardless the price. Almost all of them own house (0.86).

- Price: They are older people with average age 60. Nearly all of them own a house (0.94). They are less likely to purchase online (`store_trans = 6` while `online_trans = 3`). It is the only group that is less likely to buy online.

- Quality: The average age is 35. They are not way different with Conspicuous regarding age. But they spend much less. The percentages of male and female are similar. They prefer online shopping. More than half of them don't own a house (0.66).

- Style: They are young people with average age 24. The majority of them are female (0.81). Most of them don't own a house (0.73). They are very likely to be digital natives and prefer online shopping.

You may notice that Style group purchase more frequently online (online_trans) but the expense (online_exp) is not higher. It makes us wonder what is the average expense each time, so you have a better idea about the price range of the group.

The analytical process is aggregated instead of independent steps. The current step will shed new light on what to do next. Sometimes you need to go back to fix something in the previous steps. Let's check average one-time online and instore purchase amounts:

```
sim.dat %>%
  group_by(segment) %>%
  summarise(avg_online = round(sum(online_exp)/sum(online_trans), 2),
            avg_store = round(sum(store_exp)/sum(store_trans), 2))
```

```
## # A tibble: 4 x 3
##     segment      avg_online avg_store
##     <chr>             <dbl>     <dbl>
## 1 Conspicuous        442.       479.
## 2 Price               69.3       81.3
## 3 Quality            126.       105.
## 4 Style               92.8      121.
```

Price group has the lowest averaged one-time purchase. The Conspicuous group will pay the highest price. When we build customer profile in real life, we will also need to look at the survey summarization. You may be surprised how much information simple data manipulations can provide.

Another comman task is to check which column has missing values. It requires the program to look at each column in the data. In this case you can use `summarise_all`:

```
# apply function anyNA() to each column
# you can also assign a function vector
# such as: c("anyNA","is.factor")
dplyr::summarise_all(sim.dat, funs_(c("anyNA")))
```

```
## Warning: `funs_()` was deprecated in dplyr 0.7.0.
## Please use `funs()` instead.
## See vignette('programming') for more help
## This warning is displayed once every 8 hours.
## Call `lifecycle::last_lifecycle_warnings()` to see where this
   warning was generated.
```

```
## Warning: `funs()` was deprecated in dplyr 0.8.0.
## Please use a list of either functions or lambdas:
##
##   # Simple named list:
##   list(mean = mean, median = median)
##
##   # Auto named with `tibble::lst()`:
##   tibble::lst(mean, median)
##
##   # Using lambdas
##   list(~ mean(., trim = .2), ~ median(., na.rm = TRUE))
## This warning is displayed once every 8 hours.
## Call `lifecycle::last_lifecycle_warnings()` to see where this
   warning was generated.
```

```
##      age gender income house store_exp online_exp
## 1 FALSE  FALSE   TRUE FALSE     FALSE      FALSE
##   store_trans online_trans   Q1   Q2   Q3   Q4
```

```
## 1          FALSE           FALSE FALSE FALSE FALSE FALSE
##      Q5    Q6    Q7    Q8    Q9   Q10 segment
## 1 FALSE FALSE FALSE FALSE FALSE FALSE    FALSE
```

The above code returns a vector indicating if there is any value missing in each column.

6.1.1.4 Create New Variable

There are often situations where you need to create new variables. For example, adding online and store expense to get total expense. In this case, you will apply a function to the columns and return a column with the same length. mutate() can do it for you and append one or more new columns:

```
dplyr::mutate(sim.dat, total_exp = store_exp + online_exp)
```

The above code sums up two columns and appends the result (total_exp) to sim.dat. Another similar function is transmute(). The difference is that transmute() will delete the original columns and only keep the new ones.

```
dplyr::transmute(sim.dat, total_exp = store_exp + online_exp)
```

6.1.1.5 Merge

Similar to SQL, there are different joins in dplyr. We create two baby data sets to show how the functions work.

```
(x <- data.frame(cbind(ID = c("A", "B", "C"), x1 = c(1, 2, 3))))
```

```
##   ID x1
## 1  A  1
## 2  B  2
## 3  C  3
```

```
(y <- data.frame(cbind(ID = c("B", "C", "D"), y1 = c(T, T, F))))
```

```
##   ID    y1
## 1  B  TRUE
## 2  C  TRUE
## 3  D FALSE
```

```
# join to the left
# keep all rows in x
left_join(x, y, by = "ID")
```

```
##   ID x1   y1
## 1  A  1 <NA>
## 2  B  2 TRUE
## 3  C  3 TRUE
```

```
# get rows matched in both data sets
inner_join(x, y, by = "ID")
```

```
##   ID x1   y1
## 1  B  2 TRUE
## 2  C  3 TRUE
```

```
# get rows in either data set
full_join(x, y, by = "ID")
```

```
##   ID   x1    y1
## 1  A    1  <NA>
## 2  B    2  TRUE
## 3  C    3  TRUE
## 4  D <NA> FALSE
```

```
# filter out rows in x that can be matched in y
# it doesn't bring in any values from y
semi_join(x, y, by = "ID")
```

```
# the opposite of  semi_join()
# it gets rows in x that cannot be matched in y
# it doesn't bring in any values from y
anti_join(x, y, by = "ID")
```

There are other functions (`intersect()`, `union()` and `setdiff()`). Also the data frame version of `rbind` and `cbind` which are `bind_rows()` and `bind_col()`. We are not going to go through them all. You can try them yourself. If you understand the functions we introduced so far. It should be easy for you to figure out the rest.

6.1.2 `apply()`, `lapply()` and `sapply()` in base R

There are some powerful functions to summarize data in base R, such as `apply()`, `lapply()` and `sapply()`. They do the same basic things and are all from "apply" family: apply functions over parts of data. They differ in two important respects:

1. the type of object they apply to
2. the type of result they will return

When do we use `apply()`? When we want to apply a function to margins of an array or matrix. That means our data need to be structured. The operations can be very flexible. It returns a vector or array or list of values obtained by applying a function to margins of an array or matrix.

For example you can compute row and column sums for a matrix:

```
## simulate a matrix
x <- cbind(x1 =1:8, x2 = c(4:1, 2:5))
dimnames(x)[[1]] <- letters[1:8]
apply(x, 2, mean)
```

```
##  x1  x2
## 4.5 3.0
```

```
col.sums <- apply(x, 2, sum)
row.sums <- apply(x, 1, sum)
```

You can also apply other functions:

```
ma <- matrix(c(1:4, 1, 6:8), nrow = 2)
ma
```

```
##      [,1] [,2] [,3] [,4]
## [1,]    1    3    1    7
## [2,]    2    4    6    8
```

```
apply(ma, 1, table)   #--> a list of length 2
```

```
## [[1]]
##
## 1 3 7
## 2 1 1
##
## [[2]]
##
## 2 4 6 8
## 1 1 1 1
```

```
apply(ma, 1, stats::quantile) # 5 x n matrix with rownames
```

```
##        [,1] [,2]
## 0%        1  2.0
## 25%       1  3.5
## 50%       2  5.0
## 75%       4  6.5
## 100%      7  8.0
```

Results can have different lengths for each call. This is a trickier example. What will you get?

```
## Example with different lengths for each call
z <- array(1:24, dim = 2:4)
zseq <- apply(z, 1:2, function(x) seq_len(max(x)))
zseq            ## a 2 x 3 matrix
typeof(zseq) ## list
dim(zseq) ## 2 3
zseq[1,]
apply(z, 3, function(x) seq_len(max(x)))
```

- `lapply()` applies a function over a list, data.frame or vector and returns a list of the same length.
- `sapply()` is a user-friendly version and wrapper of `lapply()`. By default it returns a vector, matrix or if `simplify = "array"`, an array if appropriate. `apply(x, f, simplify = FALSE, USE.NAMES = FALSE)` is the same as `lapply(x, f)`. If `simplify=TRUE`, then it will return a `data.frame` instead of `list`.

Let's use some data with context to help you better understand the functions.

- Get the mean and standard deviation of all numerical variables in the dataset.

```
# Get numerical variables
sdat <- sim.dat[, lapply(sim.dat, class) %in% c("integer", "numeric")]
## Try the following code with apply() function apply(sim.dat,2,class)
## What is the problem?
```

The data frame `sdat` only includes numeric columns. Now we can go head and use `apply()` to get mean and standard deviation for each column:

```
apply(sdat, MARGIN = 2, function(x) mean(na.omit(x)))
```

```
##            age        income     store_exp     online_exp
##      3.884e+01     1.135e+05     1.357e+03      2.120e+03
##    store_trans  online_trans           Q1             Q2
##      5.350e+00     1.355e+01     3.101e+00      1.823e+00
##            Q3            Q4           Q5             Q6
##      1.992e+00     2.763e+00     2.945e+00      2.448e+00
##            Q7            Q8           Q9            Q10
##      3.434e+00     2.396e+00     3.085e+00      2.320e+00
```

Here we defined a function using `function(x) mean(na.omit(x))`. It is a very simple function. It tells R to ignore the missing value when calculating the mean. `MARGIN = 2` tells R to apply the function to each column. It is not hard to guess what `MARGIN = 1` mean. The result show that the average online expense is much higher than store expense. You can also compare the average scores across different questions. The command to calculate standard deviation is very similar. The only difference is to change `mean()` to `sd()`:

```
apply(sdat, MARGIN = 2, function(x) sd(na.omit(x)))
```

```
##            age        income     store_exp     online_exp
##         16.417     49842.287      2774.400       1731.224
##    store_trans  online_trans           Q1             Q2
```

```
##          3.696           7.957           1.450           1.168
##             Q3              Q4              Q5              Q6
##          1.402           1.155           1.284           1.439
##             Q7              Q8              Q9             Q10
##          1.456           1.154           1.118           1.136
```

Even the average online expense is higher than store expense, the standard deviation for store expense is much higher than online expense which indicates there is very likely some big/small purchase in store. We can check it quickly:

```
summary(sdat$store_exp)
```

```
##    Min. 1st Qu.  Median    Mean 3rd Qu.    Max.
##    -500     205     329    1357     597   50000
```

```
summary(sdat$online_exp)
```

```
##    Min. 1st Qu.  Median    Mean 3rd Qu.    Max.
##      69     420    1942    2120    2441    9479
```

There are some odd values in store expense. The minimum value is -500 which indicates that you should preprocess data before analyzing it. Checking those simple statistics will help you better understand your data. It then gives you some idea how to preprocess and analyze them. How about using lapply() and sapply()?

Run the following code and compare the results:

```
lapply(sdat, function(x) sd(na.omit(x)))
sapply(sdat, function(x) sd(na.omit(x)))
sapply(sdat, function(x) sd(na.omit(x)), simplify = FALSE)
```

6.2 Tidy and Reshape Data

"Tidy data" represents the information from a dataset as data
frames where each row is an observation, and each column contains
the values of a variable (i.e., an attribute of what we are observing).
Depending on the situation, the requirements on what to present
as rows and columns may change. To make data easy to work with
the problem at hand. In practice, we often need to convert data
between the "wide" and the "long" format. The process feels like
kneading the dough.

In this section, we will show how to tidy and reshape data using
tidyr packages. It is built to simplify the process of creating tidy
data. We will go through four fundamental functions:

- gather(): reshape data from wide to long
- spread(): reshape data from long to wide
- separate(): split a column into multiple columns
- unite(): combine multiple columns to one column

Take a baby subset of our exemplary clothes consumers data
to illustrate:

```
sdat<-sim.dat[1:5,1:6]
sdat
```

```
##    age gender income house store_exp online_exp
## 1   57 Female 120963   Yes     529.1      303.5
## 2   63 Female 122008   Yes     478.0      109.5
## 3   59   Male 114202   Yes     490.8      279.2
## 4   60   Male 113616   Yes     347.8      141.7
## 5   51   Male 124253   Yes     379.6      112.2
```

For the above data sdat, what if we want to reshape the data to
have a column indicating the purchasing channel (i.e. from store_exp
or online_exp) and a second column with the corresponding expense

amount? Assume we want to keep the rest of the columns the same. It is a task to change data from "wide" to "long."

```
dat_long <- tidyr::gather(sdat, "Channel","Expense",
                          store_exp, online_exp)
dat_long
```

```
##      age gender income house    Channel Expense
## 1    57 Female 120963   Yes  store_exp   529.1
## 2    63 Female 122008   Yes  store_exp   478.0
## 3    59   Male 114202   Yes  store_exp   490.8
## 4    60   Male 113616   Yes  store_exp   347.8
## 5    51   Male 124253   Yes  store_exp   379.6
## 6    57 Female 120963   Yes online_exp   303.5
## 7    63 Female 122008   Yes online_exp   109.5
## 8    59   Male 114202   Yes online_exp   279.2
## 9    60   Male 113616   Yes online_exp   141.7
## 10   51   Male 124253   Yes online_exp   112.2
```

The above code gathers two variables (store_exp and online_exp), and collapses them into key-value pairs (Channel and Expense), duplicating all other columns as needed.

You can run a regression to study the effect of purchasing channel as follows:

```
# Here we use all observations from sim.dat
# Don't show result here

msdat <- tidyr::gather(sim.dat[, 1:6], "Channel","Expense",
                       store_exp, online_exp)
fit <- lm(Expense ~ gender + house + income + Channel + age,
          data = msdat)
summary(fit)
```

Sometimes we want to reshape the data from "long" to "wide."
For example, you want to compare the online and in-store expense
between male and female based on house ownership.

We need to reshape the wide data frame `dat_wide` to a long
format by spreading the key-value pairs across multiple columns.
And then summarize the long data frame `dat_long`, grouping by`house`
and `gender`.

```
dat_wide = tidyr::spread(dat_long, Channel, Expense)
# you can check what dat_long is like
dat_wide %>%
  dplyr::group_by(house, gender) %>%
  dplyr::summarise(total_online_exp = sum(online_exp),
                   total_store_exp = sum(store_exp))
```

```
## # A tibble: 2 x 4
## # Groups:   house [1]
##   house gender total_online_exp total_store_exp
##   <chr> <chr>             <dbl>           <dbl>
## 1 Yes   Female             413.           1007.
## 2 Yes   Male               533.           1218.
```

The above code also uses the functions in the `dplyr` package
introduced in the previous section. Here we use `package::function`
to make clear the package name. It is not necessary if the package
is already loaded.

Another pair of functions that do opposite manipulations are
`separate()` and `unite()`.

```
sepdat<- dat_long %>%
  separate(Channel, c("Source", "Type"))
sepdat
```

```
##   age gender income house Source Type Expense
## 1  57 Female 120963   Yes  store  exp   529.1
```

```
## 2    63 Female 122008    Yes   store   exp   478.0
## 3    59   Male 114202    Yes   store   exp   490.8
## 4    60   Male 113616    Yes   store   exp   347.8
## 5    51   Male 124253    Yes   store   exp   379.6
## 6    57 Female 120963    Yes  online   exp   303.5
## 7    63 Female 122008    Yes  online   exp   109.5
## 8    59   Male 114202    Yes  online   exp   279.2
## 9    60   Male 113616    Yes  online   exp   141.7
## 10   51   Male 124253    Yes  online   exp   112.2
```

You can see that the function separates the original column "Channel" to two new columns "Source" and "Type". You can use sep = to set the string or regular expression to separate the column. By default, it is "_".

The unite() function will do the opposite: combining two columns. It is the generalization of paste() to a data frame.

```
sepdat %>%
  unite("Channel", Source, Type, sep = "_")
```

```
##      age gender income house    Channel Expense
## 1    57 Female 120963    Yes  store_exp   529.1
## 2    63 Female 122008    Yes  store_exp   478.0
## 3    59   Male 114202    Yes  store_exp   490.8
## 4    60   Male 113616    Yes  store_exp   347.8
## 5    51   Male 124253    Yes  store_exp   379.6
## 6    57 Female 120963    Yes online_exp   303.5
## 7    63 Female 122008    Yes online_exp   109.5
## 8    59   Male 114202    Yes online_exp   279.2
## 9    60   Male 113616    Yes online_exp   141.7
## 10   51   Male 124253    Yes online_exp   112.2
```

The reshaping manipulations may be the trickiest part. You have to practice a lot to get familiar with those functions. Unfortunately, there is no shortcut.

7

Model Tuning Strategy

When training a machine learning model, there are many decisions to make. For example, when training a random forest, you need to decide the number of trees and the number of variables to use at each node. For the lasso method, you need to determine the penalty parameter. Unlike the parameters derived by training (such as the coefficients in a linear regression model), those parameters are used to control the learning process and are called hyperparameters. To train a model, you need to set the value of hyperparameters.

A common way to make those decisions is to split the data into training and testing sets. Use training data to fit models with different parameter values and apply the fitted models to the testing data. And then, find the hyperparameter value that gives the best testing performance. Data splitting is also used in model selection and evaluation, where you access the correctness of a model on an evaluation set and compare different models to find the best one.

In practice, applying machine learning is a highly iterative process. This chapter will illustrate the practical aspects of model tuning. We will talk about different types of model error, source of model error, hyperparameter tuning, how to set up your data, and how to ensure your model implementation is correct (i.e. model selection and evalutaion).

Load the R packages first:

```
# install packages from CRAN
p_needed <- c('ggplot2','tidyr', 'caret', 'dplyr',
              'lattice', 'proxy', 'caret')
packages <- rownames(installed.packages())
p_to_install <- p_needed[!(p_needed %in% packages)]
if (length(p_to_install) > 0) {
```

DOI: 10.1201/9781351132916-7

```
    install.packages(p_to_install)
}

lapply(p_needed, require, character.only = TRUE)
```

7.1 Variance-Bias Trade-Off

Assume $\mathbf{X}$ is $n \times p$ observation matrix and $\mathbf{y}$ is response variable, we have:

$$\mathbf{y} = f(\mathbf{X}) + \epsilon \qquad (7.1)$$

where ϵ is the random error with a mean of zero. The function $f(\cdot)$ is our modeling target, which represents the information in the response variable that predictors can explain. The main goal of estimating $f(\cdot)$ is inference or prediction, or sometimes both. In general, there is a trade-off between flexibility and interpretability of the model. So data scientists need to comprehend the delicate balance between these two.

Depending on the modeling purposes, the requirement for interpretability varies. If the prediction is the only goal, then as long as the prediction is accurate enough, the interpretability is not under consideration. In this case, people can use "black box" model, such as random forest, boosting tree, neural network and so on. These models are very flexible but usually difficult to explain. Their accuracy is usually higher on the training set, but not necessary when it predicts. It is not surprising since those models have a huge number of parameters and high flexibility that they can "memorize" the entire training data. A paper by Chiyuan Zhang et al. in 2017 pointed out that "Deep neural networks (even just two-layer net) easily fit random labels" (Zhang et al., 2017). The traditional forms of regularization, such as weight decay, dropout, and data augmentation, fail to control generalization error. It poses a conceptual challenge to statistical theory and also calls our attention when we use such black-box models.

There are two kinds of application problems: complete information problem and incomplete information problem. The complete
information problem has all the information you need to know the
correct response. Take the famous cat recognition, for example, all
the information you need to identify a cat is in the picture. In this
situation, the algorithm that penetrates the data the most wins.
There are some other similar problems such as the self-driving car,
chess game, facial recognition and speech recognition. But in most
of the data science applications, the information is incomplete. If
you want to know whether a customer is going to purchase again or
not, it is unlikely to have 360-degree of the customer's information.
You may have their historical purchasing record, discounts and
service received. But you don't know if the customer sees your advertisement, or has a friend recommends competitor's product, or
encounters some unhappy purchasing experience somewhere. There
could be a myriad of factors that will influence the customer's
purchase decision while what you have as data is only a small
part. To make things worse, in many cases, you don't even know
what you don't know. Deep learning does not have much advantage
in solving these problems, especially when the size of the data is
relatively small. Instead, some parametric models often work better
in this situation. You will comprehend this more after learning the
different types of model error.

Assume we have $\hat{f}$ which is an estimator of f. Then we can
further get $\hat{\mathbf{y}} = \hat{f}(\mathbf{X})$. The predicted error is divided into two parts,
systematic error, and random error:

$$
\begin{aligned}
E(\mathbf{y} - \hat{\mathbf{y}})^2 &= E[f(\mathbf{X}) + \epsilon - \hat{f}(\mathbf{X})]^2 \\
&= \underbrace{E[f(\mathbf{X}) - \hat{f}(\mathbf{X})]^2}_{(1)} + \underbrace{Var(\epsilon)}_{(2)}
\end{aligned} \tag{7.2}
$$

It is also called Mean Square Error (MSE) where (1) is the
systematic error. It exists because $\hat{f}$ usually does not entirely
describe the "systematic relation" between X and y which refers
to the stable relationship that exists across different samples or
time. Model improvement can help reduce this kind of error; (2)
is the random error which represents the part of y that cannot

be explained by X. A more complex model does not reduce the random error. There are three reasons for random error:

1. The current sample is not representative, so the pattern in one sample set does not generalize to a broader scale.
2. The information is incomplete. In other words, you don't have all variables needed to explain the response.
3. There is measurement error in the variables.

Deep learning has significant success solving problems with complete information and usually with low measurement error. As mentioned before, in a task like image recognition, all you need are the pixels in the pictures. So in deep learning applications, increasing the sample size can improve the model performance significantly. But it may not perform well in problems with incomplete information. The biggest problem with the black-box model is that it fits random error, i.e., over-fitting. The notable feature of random error is that it varies over different samples. So one way to determine whether overfitting happens is to reserve a part of the data as the test set and then check the performance of the trained model on the test data. Note that overfitting is a general problem from which any model could suffer. However, since black-box models usually have a large number of parameters, it is much more susceptible to over-fitting.

The systematic error $E[f(\mathbf{X}) - \hat{f}(\mathbf{X})]^2$ can be further decomposed as:

$$\left(f(\mathbf{X}) - E[\hat{f}(\mathbf{X})] + E[\hat{f}(\mathbf{X})] - \hat{f}(\mathbf{X})\right)$$
$$= E\left(E[\hat{f}(\mathbf{X})] - f(\mathbf{X})\right)^2 + E\left(\hat{f}(\mathbf{X}) - E[\hat{f}(\mathbf{X})]\right)^2$$
$$= [Bias(\hat{f}(\mathbf{X}))]^2 + Var(\hat{f}(\mathbf{X})) \tag{7.3}$$

The systematic error consists of two parts, $Bias(\hat{f}(\mathbf{X}))$ and $Var(\hat{f}(\mathbf{X}))$. To minimize the systematic error, we need to minimize both. The bias represents the error caused by the model's approximation of the reality, i.e., systematic relation, which may be very complex. For example, linear regression assumes a linear relationship between the predictors and the response, but rarely is there a perfect linear relationship in real life. So linear regression

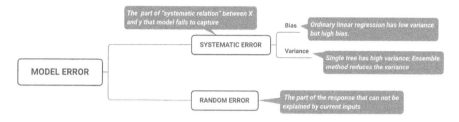

FIGURE 7.1
Types of model error

is more likely to have a high bias. Generally, the more flexible the model is, the higher the variance. However, this does not guarantee that complex models will outperform simpler ones, such as linear regression. If the real relationship f is linear, then linear regression is unbiased. It is difficult for a more flexible model to compete. An ideal learning method has low variance and bias. However, it is easy to find a model with a low bias but high variance (by fitting a tree) or a method with a low variance but high bias (by fitting a straight line). That is why we call it a trade-off.

To explore bias and variance, let's begin with a simple simulation. We will simulate data with a non-linear relationship and fit different models using the simulated data. An intuitive way to show is to compare the plots of various models.

The code below simulates one predictor (x) and one response variable (fx). The relationship between x and fx is non-linear. You need to load the multiplot function by running source('http://bit.ly/2KeEIg9'). The function assembles multiple plots on a canvas.

```
source('http://bit.ly/2KeEIg9')
# randomly simulate some non-linear samples
x = seq(1, 10, 0.01) * pi
e = rnorm(length(x), mean = 0, sd = 0.2)
fx <- sin(x) + e + sqrt(x)
dat = data.frame(x, fx)
```

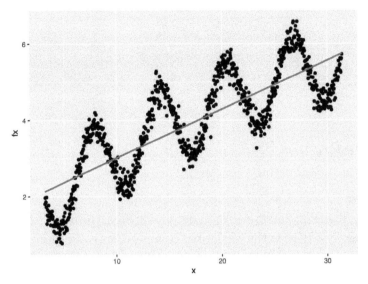

FIGURE 7.2
High bias model

Then fit a linear regression on the data:

```
# plot fitting result
library(ggplot2)
ggplot(dat, aes(x, fx)) +
    geom_point() +
    geom_smooth(method = "lm", se = FALSE)
```

```
## `geom_smooth()` using formula 'y ~ x'
```

Despite a large sample size, trained linear regression cannot describe the relationship very well. In other words, in this case, the model has a high bias (figure 7.2). It is also called underfitting.

Since the estimated parameters will be somewhat different for different samples, there is a variance in estimates. Intuitively, it gives you some sense of the extent to which the estimates would change if we fit the same model with different samples (presumably, they are from the same population). Ideally, the change is small. For high variance models, small changes in the training data

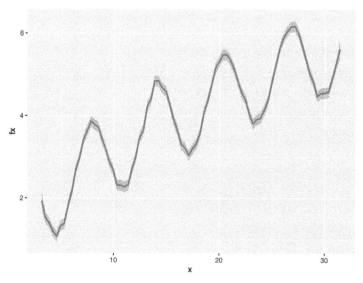

FIGURE 7.3
High variance model

result in very different estimates. Generally, a model with high flexibility also has high variance, such as the CART tree and the initial boosting method. To overcome that problem, the Random Forest and Gradient Boosting Model aim to reduce the variance by summarizing the results obtained from different samples.

Let's fit the above data using a smoothing method that is highly flexible and can fit the current data tightly:

```
ggplot(dat, aes(x, fx)) + geom_smooth(span = 0.03)
```

The resulting plot (figure 7.3) indicates the smoothing method fit the data much better and it has a much smaller bias. However, this method has a high variance. If we simulate different subsets of the sample, the result curve will change significantly:

```
# set random seed
set.seed(2016)
```

```r
# sample part of the data to fit model sample 1
idx1 = sample(1:length(x), 100)
dat1 = data.frame(x1 = x[idx1], fx1 = fx[idx1])
p1 = ggplot(dat1, aes(x1, fx1)) +
  geom_smooth(span = 0.03) +
  geom_point()

# sample 2
idx2 = sample(1:length(x), 100)
dat2 = data.frame(x2 = x[idx2], fx2 = fx[idx2])
p2 = ggplot(dat2, aes(x2, fx2)) +
  geom_smooth(span = 0.03) +
  geom_point()

# sample 3
idx3 = sample(1:length(x), 100)
dat3 = data.frame(x3 = x[idx3], fx3 = fx[idx3])
p3 = ggplot(dat3, aes(x3, fx3)) +
  geom_smooth(span = 0.03) +
  geom_point()

# sample 4
idx4 = sample(1:length(x), 100)
dat4 = data.frame(x4 = x[idx4], fx4 = fx[idx4])
p4 = ggplot(dat4, aes(x4, fx4)) +
  geom_smooth(span = 0.03) +
  geom_point()

multiplot(p1, p2, p3, p4, cols = 2)
```

The fitted lines (blue) change over different samples which means it has high variance. People also call it overfitting. Fitting the linear model using the same four subsets, the result barely changes:

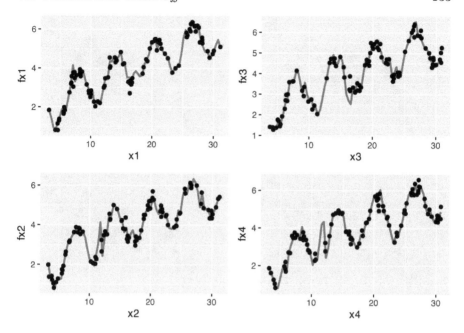

```
p1 = ggplot(dat1, aes(x1, fx1)) +
  geom_smooth(method = "lm", se = FALSE) +
  geom_point()

p2 = ggplot(dat2, aes(x2, fx2)) +
  geom_smooth(method = "lm", se = FALSE) +
  geom_point()

p3 = ggplot(dat3, aes(x3, fx3)) +
  geom_smooth(method = "lm", se = FALSE) +
  geom_point()

p4 = ggplot(dat4, aes(x4, fx4)) +
  geom_smooth(method = "lm", se = FALSE) +
  geom_point()

multiplot(p1, p2, p3, p4, cols = 2)
```

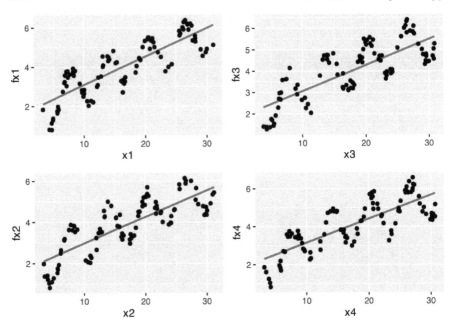

In general, the variance $(Var(\hat{f}(\mathbf{X})))$ **increases** and the bias $(Bias(\hat{f}(\mathbf{X})))$ **decreases** as the model flexibility increases. Variance and bias together determine the systematic error. As we increase the flexibility of the model, at first the rate at which $Bias(\hat{f}(\mathbf{X}))$ decreases is faster than $Var(\hat{f}(\mathbf{X}))$, so the MSE decreases. However, to some degree, higher flexibility has little effect on $Bias(\hat{f}(\mathbf{X}))$ but $Var(\hat{f}(\mathbf{X}))$ increases significantly, so the MSE increases. A typical criterion to balance bias and variance is to choose a model has the minimum MSE as described with detail in next section.

7.2 Data Splitting and Resampling

Highly adaptable models can model complex relationships. However, they tend to overfit, which leads to a poor prediction by learning too much from the current sample set. Those models are susceptible to the specific sample set used to fit them. The model prediction may be off when future data is unlike past data. Conversely, a simple model, such as ordinary linear regression, tends to underfit, leading to a poor prediction by learning too little from the data. It systematically over-predicts or under-predicts the data regardless of how well future data resemble past data.

Model evaluation is essential to assess the efficacy of a model. A modeler needs to understand how a model fits the existing data and how it would work on future data. Also, trying multiple models and comparing them is always a good practice. All these need data splitting and resampling.

7.2.1 Data Splitting

Data splitting is to put part of the data aside as an evaluation set (or hold-outs, out-of-bag samples) and use the rest for model tuning. Training samples are also called in-sample. Model performance metrics evaluated using in-sample are retrodictive, not predictive.

Traditional business intelligence usually handles data description. Answer simple questions by querying and summarizing the data, such as:

- What are the monthly sales of a product in 2020?
- What is the number of site visits in the past month?
- What is the sales difference in 2021 for two different product designs?

There is no need to go through the tedious process of splitting the data, tuning, and evaluating a model to answer questions of this kind.

Some models have hyperparameters (aka. tuning parameters) not derived by training the model, such as the penalty parameter in lasso, the number of trees in a random forest, and the learning rate in deep learning. They often control the model's process, and no analytical formula is available to calculate the optimized value. A poor choice can result in over-fitting, under-fitting, or optimization failure. A standard approach to searching for the best tuning parameters is through cross-validation, which is a data resampling approach.

To get a reasonable performance precision based on a single test set, the size of the test set may need to be large. So a conventional approach is to use a subset of samples to fit the model and use the rest to evaluate model performance. This process will repeat multiple times to get a performance profile. In that sense, resampling is based on splitting. The general steps are

---

Algorithm 1 General resampling steps

1: Define a set of candidate values for tuning parameter(s)
2: Resample data
3: **for** Each candidate value in the set **do**
4:     Fit model
5:     Predict hold-out
6:     Calculate performance
7: **end for**
8: Aggregate the results
9: Determine the final tuning parameter
10: Refit the model with the entire data set

The above is an outline of the general procedure to tune parameters. Now let's focus on the critical part of the process: data splitting. Ideally, we should evaluate the model using samples not used to build or fine-tune the model. So it provides an unbiased sense of model effectiveness. When the sample size is large, it is a good practice to set aside part of the samples to evaluate the final model. People use **training data** to indicate the sample set used to fit the model. Use **testing data** to tune hyperparameters and **validation data** to evaluate performance and compare different models.

Let's focus on data splitting in the model tuning process, where we split data into training and testing sets.

The first decision is the proportion of data in the test set. There are two factors to consider here: (1) sample size; (2) computation intensity. Suppose the sample size is large enough, which is the most common situation according to my experience. In that case, you can try using 20%, 30%, and 40% of the data as the test set and see which works best. If the model is computationally intense, you may consider starting from a smaller subset to train the model and hence have a higher portion of data in the test set. You may need to increase the training set depending on how it performs. If

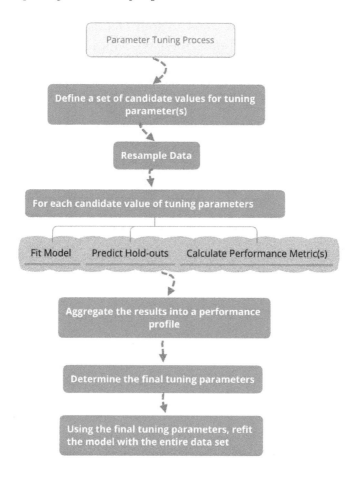

FIGURE 7.4
Parameter tuning process

the sample size is small, you can use cross-validation or bootstrap, which is the topic of the next section.

The next is to decide which samples are in the test set. There is a desire to make the training and test sets as similar as possible. A simple way is to split data randomly, which does not control for any data attributes. However, sometimes we may want to ensure that training and testing data have a similar outcome distribution. For example, suppose you want to predict the likelihood of customer retention. In that case, you want two data sets with a similar percentage of retained customers.

There are three main ways to split the data that account for the similarity of resulted data sets. We will describe the three approaches using the clothing company's customer data as examples.

(1) Split data according to the outcome variable

Assume the outcome variable is customer segment (column segment), and we decide to use 80% as training and 20% as testing. The goal is to make the proportions of the categories in the two sets as similar as possible. The createDataPartition() function in caret will return a balanced splitting based on assigned variable.

```
# load data
sim.dat <- read.csv("http://bit.ly/2P5gTw4")
library(caret)
# set random seed to make sure reproducibility
set.seed(3456)
trainIndex <- createDataPartition(sim.dat$segment,
                                  p = 0.8,
                                  list = FALSE,
                                  times = 1)
head(trainIndex)
```

```
##         Resample1
## [1,]            1
## [2,]            2
## [3,]            3
## [4,]            4
## [5,]            6
## [6,]            7
```

The list = FALSE in the call to createDataPartition is to return a data frame. The times = 1 tells R how many times you want to split the data. Here we only do it once, but you can repeat the splitting multiple times. In that case, the function will return multiple vectors indicating the rows to training/test. You can set times = 2 and rerun the above code to see the result. Then we can use the returned indicator vector `trainIndex` to get training and test sets:

```
# get training set
datTrain <- sim.dat[trainIndex, ]
# get test set
datTest <- sim.dat[-trainIndex, ]
```

According to the setting, there are 800 samples in the training set and 200 in the testing set. Let's check the distribution of the two groups:

```
datTrain %>%
  dplyr::group_by(segment) %>%
  dplyr::summarise(count = n(),
          percentage = round(length(segment)/nrow(datTrain), 2))
```

```
## # A tibble: 4 x 3
##    segment      count percentage
##    <chr>        <int>      <dbl>
## 1 Conspicuous    160        0.2
## 2 Price          200        0.25
## 3 Quality        160        0.2
## 4 Style          280        0.35
```

```
datTest %>%
  dplyr::group_by(segment) %>%
  dplyr::summarise(count = n(),
          percentage = round(length(segment)/nrow(datTest), 2))
```

```
## # A tibble: 4 x 3
##    segment      count percentage
##    <chr>        <int>      <dbl>
```

```
## 1 Conspicuous      40         0.2
## 2 Price            50         0.25
## 3 Quality          40         0.2
## 4 Style            70         0.35
```

The percentages are the same for these two sets. In practice, it is possible that the distributions are not identical but should be close.

(2)   Divide data according to predictors

An alternative way is to split data based on the predictors. The goal is to get a diverse subset from a dataset to represent the sample. In other words, we need an algorithm to identify the n most diverse samples from a dataset with size N. However, the task is generally infeasible for non-trivial values of n and N (Willett, 2004). And hence practicable approaches to dissimilarity-based selection involve approximate methods that are sub-optimal. A major class of algorithms split the data on *maximum dissimilarity sampling*. The process starts from:

- Initialize a single sample as starting test set
- Calculate the dissimilarity between this initial sample and each remaining samples in the dataset
- Add the most dissimilar unallocated sample to the test set

To move forward, we need to define the dissimilarity between groups. Each definition results in a different version of the algorithm and hence a different subset. It is the same problem as in hierarchical clustering where you need to define a way to measure the distance between clusters. The possible approaches are to use minimum, maximum, sum of all distances, the average of all distances, etc. Unfortunately, there is not a single best choice, and you may have to try multiple methods and check the resulted sample sets. R users can implement the algorithm using `maxDissim()` function from `caret` package. The `obj` argument is to set the definition of dissimilarity. Refer to the help documentation for more details (`?maxDissim`).

Let's use two variables (`age` and `income`) from the customer data as an example to illustrate how it works in R and compare maximum dissimilarity sampling with random sampling.

```
library(lattice)
# select variables
testing <- subset(sim.dat, select = c("age", "income"))
```

Random select 5 samples as initial subset (`start`), the rest will be in `samplePool`:

```
set.seed(5)
# select 5 random samples
startSet <- sample(1:dim(testing)[1], 5)
start <- testing[startSet, ]
# save the rest in data frame 'samplePool'
samplePool <- testing[-startSet, ]
```

Use `maxDissim()` to select another 5 samples from `samplePool` that are as different as possible with the initical set `start`:

```
selectId <- maxDissim(start, samplePool, obj = minDiss, n = 5)
minDissSet <- samplePool[selectId, ]
```

The `obj = minDiss` in the above code tells R to use minimum dissimilarity to define the distance between groups. Next, random select 5 samples from `samplePool` in data frame `RandomSet`:

```
selectId <- sample(1:dim(samplePool)[1], 5)
RandomSet <- samplePool[selectId, ]
```

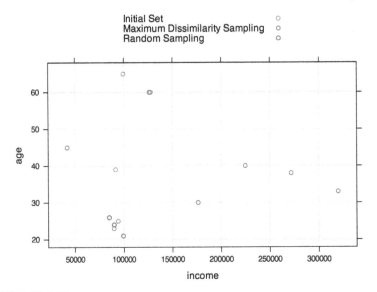

FIGURE 7.5

Compare maximum dissimilarity sampling with random sampling

Plot the resulted set to compare different sampling methods:

```
start$group <- rep("Initial Set", nrow(start))
minDissSet$group <- rep("Maximum Dissimilarity Sampling",
                        nrow(minDissSet))
RandomSet$group <- rep("Random Sampling",
                        nrow(RandomSet))
xyplot(age ~ income,
       data = rbind(start, minDissSet, RandomSet),
       grid = TRUE,
       group = group,
       auto.key = TRUE)
```

The points from maximum dissimilarity sampling are far away from the initial samples (figure 7.5, while the random samples are much closer to the initial ones. Why do we need a diverse subset? Because we hope the test set to be representative. If all test set samples are from respondents younger than 30, model performance on the test set has a high risk to fail to tell you how the model will perform on more general population.

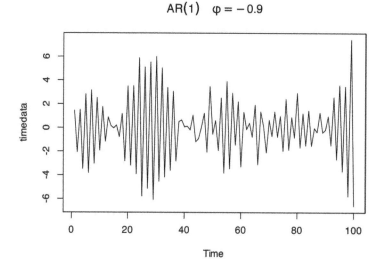

FIGURE 7.6
Divide data according to time

- Divide data according to time

For time series data, random sampling is usually not the best way. There is an approach to divide data according to time-series. Since time series is beyond the scope of this book, there is not much discussion here. For more detail of this method, see (Hyndman and Athanasopoulos, 2013). We will use a simulated first-order autoregressive model (i.e., AR(1) model) time-series data with 100 observations to show how to implement using the function `createTimeSlices()` in the `caret` package.

```
# simulte AR(1) time series samples
timedata = arima.sim(list(order=c(1,0,0), ar=-.9), n=100)
# plot time series
plot(timedata, main=(expression(AR(1)~~~phi==-.9)))
```

Figure 7.6 shows 100 simulated time series observation. The goal is to make sure both training and test set to cover the whole period.

```
timeSlices <- createTimeSlices(1:length(timedata),
                     initialWindow = 36,
                     horizon = 12,
                     fixedWindow = T)
str(timeSlices,max.level = 1)
```

```
## List of 2
##  $ train:List of 53
##  $ test :List of 53
```

There are three arguments in the above `createTimeSlices()`.

- `initialWindow`: The initial number of consecutive values in each training set sample
- `horizon`: the number of consecutive values in test set sample
- `fixedWindow`: if FALSE, all training samples start at 1

The function returns two lists, one for the training set, the other for the test set. Let's look at the first training sample:

```
# get result for the 1st training set
trainSlices <- timeSlices[[1]]
# get result for the 1st test set
testSlices <- timeSlices[[2]]
# check the index for the 1st training and test set
trainSlices[[1]]
```

```
##  [1]  1  2  3  4  5  6  7  8  9 10 11 12 13 14 15 16 17
## [18] 18 19 20 21 22 23 24 25 26 27 28 29 30 31 32 33 34
## [35] 35 36
```

```
testSlices[[1]]
```

```
##  [1] 37 38 39 40 41 42 43 44 45 46 47 48
```

The first training set is consist of sample 1-36 in the dataset (`initialWindow = 36`). Then sample 37-48 are in the first test set (`horizon = 12`). Type `head(trainSlices)` or `head(testSlices)` to check the later samples. If you are not clear about the argument `fixedWindow`, try to change the setting to be `F` and check the change in `trainSlices` and `testSlices`.

Understand and implement data splitting is not difficult. But there are two things to note:

1. The randomness in the splitting process will lead to uncertainty in performance measurement.
2. When the dataset is small, it can be too expensive to leave out test set. In this situation, if collecting more data is just not possible, the best shot is to use leave-one-out cross-validation which is discussed in the next section.

7.2.2 Resampling

You can consider resampling as repeated splitting. The basic idea is: use part of the data to fit model and then use the rest of data to calculate model performance. Repeat the process multiple times and aggregate the results. The differences in resampling techniques usually center around the ways to choose subsamples. There are two main reasons that we may need resampling:

1. Estimate tuning parameters through resampling. Some examples of models with such parameters are Support Vector Machine (SVM), models including the penalty (LASSO) and random forest.

2. For models without tuning parameter, such as ordinary linear regression and partial least square regression, the model fitting doesn't require resampling. But you can study the model stability through resampling.

We will introduce three most common resampling techniques: k-fold cross-validation, repeated training/test splitting, and bootstrap.

7.2.2.1   k-fold Cross-Validation

k-fold cross-validation is to partition the original sample into k equal size subsamples (folds). Use one of the k folds to validate the model and the rest $k-1$ to train model. Then repeat the process k times with each of the k folds as the test set. Aggregate the results into a performance profile.

Denote by $\hat{f}^{-\kappa}(X)$ the fitted function, computed with the κth fold removed and x_i^κ the predictors for samples in left-out fold. The process of k-fold cross-validation is as follows:

Algorithm 2 k-fold cross-validation

1: Partition the original sample into k equal size folds
2: **for** $\kappa = 1...k$ **do**
3:     Use data other than fold κ to train the model $\hat{f}^{-\kappa}(X)$
4:     Apply $\hat{f}^{-\kappa}(X)$ to predict fold κ to get $\hat{f}^{-\kappa}(x_i^\kappa)$
5: **end for**
6: Aggregate the results

$$\hat{Error} = \frac{1}{N}\Sigma_{\kappa=1}^k \Sigma_{x_i^\kappa} L(y_i^\kappa, \hat{f}^{-\kappa}(x_i^\kappa))$$

It is a standard way to find the value of tuning parameter that gives you the best performance. It is also a way to study the variability of model performance.

Figure 7.7 represents a 5-fold cross-validation example.

A special case of k-fold cross-validation is Leave One Out Cross Validation (LOOCV) where $k = 1$. When sample size is small, it is desired to use as many data to train the model. Most of the functions have default setting $k = 10$. The choice is usually 5–10 in practice, but there is no standard rule. The more folds to use, the more samples are used to fit model, and then the performance estimate is closer to the theoretical performance. Meanwhile, the variance of the performance is larger since the samples to fit model in different iterations are more similar. However, LOOCV has high computational cost since the number of interactions is the same as the sample size and each model fit uses a subset that is nearly the

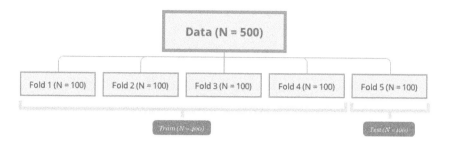

FIGURE 7.7
5-fold cross-validation

same size of the training set. On the other hand, when k is small (such as 2 or 3), the computation is more efficient, but the bias will increase. When the sample size is large, the impact of k becomes marginal.

Chapter 7 of (Hastie et al., 2008) presents a more in-depth and more detailed discussion about the bias-variance trade-off in k-fold cross-validation.

You can implement k-fold cross-validation using `createFolds()` in `caret`:

```
library(caret)
class <- sim.dat$segment
# creat k-folds
set.seed(1)
cv <- createFolds(class, k = 10, returnTrain = T)
str(cv)
```

```
## List of 10
##  $ Fold01: int [1:900] 1 2 3 4 5 6 7 8 9 10 ...
##  $ Fold02: int [1:900] 1 2 3 4 5 6 7 9 10 11 ...
##  $ Fold03: int [1:900] 1 2 3 4 5 6 7 8 10 11 ...
##  $ Fold04: int [1:900] 1 2 3 4 5 6 7 8 9 11 ...
##  $ Fold05: int [1:900] 1 3 4 6 7 8 9 10 11 12 ...
##  $ Fold06: int [1:900] 1 2 3 4 5 6 7 8 9 10 ...
##  $ Fold07: int [1:900] 2 3 4 5 6 7 8 9 10 11 ...
```

```
##   $ Fold08: int [1:900] 1 2 3 4 5 8 9 10 11 12 ...
##   $ Fold09: int [1:900] 1 2 4 5 6 7 8 9 10 11 ...
##   $ Fold10: int [1:900] 1 2 3 5 6 7 8 9 10 11 ...
```

The above code creates ten folds (k=10) according to the customer segments (we set class to be the categorical variable segment). The function returns a list of 10 with the index of rows in training set.

7.2.2.2 Repeated Training/Test Splits

In fact, this method is nothing but repeating the training/test set division on the original data. Fit the model with the training set, and evaluate the model with the test set. Unlike k-fold cross-validation, the test set generated by this procedure may have duplicate samples. A sample usually shows up in more than one test sets. There is no standard rule for split ratio and number of repetitions. The most common choice in practice is to use 75% to 80% of the total sample for training. The remaining samples are for validation. The more sample in the training set, the less biased the model performance estimate is. Increasing the repetitions can reduce the uncertainty in the performance estimates. Of course, it is at the cost of computational time when the model is complex. The number of repetitions is also related to the sample size of the test set. If the size is small, the performance estimate is more volatile. In this case, the number of repetitions needs to be higher to deal with the uncertainty of the evaluation results.

We can use the same function (createDataPartition ()) as before. If you look back, you will see times = 1. The only thing to change is to set it to the number of repetitions.

```
trainIndex <- createDataPartition(sim.dat$segment,
                                  p = .8,
                                  list = FALSE,
                                  times = 5)
dplyr::glimpse(trainIndex)
```

```
##   int [1:800, 1:5] 1 3 4 5 6 7 8 9 10 11 ...
##   - attr(*, "dimnames")=List of 2
##   ..$ : NULL
##   ..$ : chr [1:5] "Resample1" "Resample2" "Resample3" "Resample4" ...
```

Once know how to split the data, the repetition comes naturally.

7.2.2.3 Bootstrap Methods

Bootstrap is a powerful statistical tool (a little magic too). It can be used to analyze the uncertainty of parameter estimates (Efron and Tibshirani, 1986) quantitatively. For example, estimate the standard deviation of linear regression coefficients. The power of this method is that the concept is so simple that it can be easily applied to any model as long as the computation allows. However, you can hardly obtain the standard deviation for some models by using the traditional statistical inference.

Since it is with replacement, a sample can be selected multiple times, and the bootstrap sample size is the same as the original data. So for every bootstrap set, there are some left-out samples, which is also called "out-of-bag samples." The out-of-bag sample is used to evaluate the model. Efron points out that under normal circumstances (Efron, 1983), bootstrap estimates the error rate of the model with more certainty. The probability of an observation i in bootstrap sample B is:

$$Pri \in B = 1 - \left(1 - \frac{1}{N}\right)^N$$
$$\approx 1 - e^{-1}$$
$$= 0.632$$

On average, 63.2% of the observations appeared at least once in a bootstrap sample, so the estimation bias is similar to 2-fold cross-validation. As mentioned earlier, the smaller the number of folds, the larger the bias. Increasing the sample size will ease the problem. In general, bootstrap has larger bias and smaller variance than cross-validation. Efron came up the following ".632 estimator" to alleviate this bias:

$$(0.632 \times original\ bootstrap\ estimate)$$
$$+ (0.368 \times apparent\ error\ rate)$$

The apparent error rate is the error rate when the data is used twice, both to fit the model and to check its accuracy and it is apparently over-optimistic. The modified bootstrap estimate reduces the bias but can be unstable with small samples size. This estimate can also be unduly optimistic when the model severely over-fits since the apparent error rate will be close to zero. Efron and Tibshirani (Efron and Tibshirani, 1997) discuss another technique, called the "632+ method," for adjusting the bootstrap estimates.

8

Measuring Performance

To compare different models, we need a way to measure model performance. There are various metrics to use. To better understand the strengths and weaknesses of a model, you need to look at it through multiple metrics. In this chapter, we will introduce some of the most common performance measurement metrics.

Load the R packages first:

```
# install packages from CRAN
p_needed <- c('caret', 'dplyr', 'randomForest',
              'readr', 'car', 'pROC', 'fmsb', 'caret')

packages <- rownames(installed.packages())
p_to_install <- p_needed[!(p_needed %in% packages)]

if (length(p_to_install) > 0) {
    install.packages(p_to_install)
}

lapply(p_needed, require, character.only = TRUE)
```

8.1 Regression Model Performance

Mean Squared Error (MSE) measures the average of the squares of the errors – that is, the average squared difference between the estimated values ($\hat{y}_i$) and the actual value (y_i). The Root Mean Squared Error (RMSE) is the root square of the MSE.

DOI: 10.1201/9781351132916-8

$$MSE = \frac{1}{n} \sum_{i=1}^{n} (y_i - \hat{y}_i)^2$$

$$RMSE = \sqrt{\frac{1}{n} \sum_{i=1}^{n} (y_i - \hat{y}_i)^2}$$

Both are the common measurements for the regression model performance. Let's use the previous income prediction as an example. Fit a simple linear model:

```
sim.dat <- read.csv("http://bit.ly/2P5gTw4")
fit<- lm(formula = income ~ store_exp + online_exp + store_trans +
    online_trans, data = sim.dat)
summary(fit)
```

```
##
## Call:
## lm(formula = income ~ store_exp + online_exp + store_trans +
##     online_trans, data = sim.dat)
##
## Residuals:
##      Min      1Q  Median      3Q     Max
## -128768  -15804     441   13375  150945
##
## Coefficients:
##                 Estimate Std. Error t value Pr(>|t|)
## (Intercept)    85711.680   3651.599   23.47  < 2e-16 ***
## store_exp          3.198      0.475    6.73  3.3e-11 ***
## online_exp         8.995      0.894   10.06  < 2e-16 ***
## store_trans     4631.751    436.478   10.61  < 2e-16 ***
## online_trans   -1451.162    178.835   -8.11  1.8e-15 ***
## ---
## Signif. codes:
## 0 '***' 0.001 '**' 0.01 '*' 0.05 '.' 0.1 ' ' 1
##
## Residual standard error: 31500 on 811 degrees of freedom
##    (184 observations deleted due to missingness)
```

```
## Multiple R-squared:  0.602,  Adjusted R-squared:   0.6
## F-statistic:  306 on 4 and 811 DF,  p-value: <2e-16
```

You can calculate the RMSE:

```
y <- sim.dat$income
yhat <- predict(fit, sim.dat)
MSE <- mean((y - yhat)^2, na.rm = T )
RMSE <- sqrt(MSE)
RMSE
```

```
## [1] 31433
```

Another common performance measure for the regression model is R-Squared, often denoted as R^2. It is the square of the correlation between the fitted value and the observed value. It is often explained as the percentage of the information in the data that the model can explain. The above model returns an R-squared $= 0.602$, which indicates the model can explain 60.2% of the variance in variable income. While R^2 is easy to explain, it is not a direct measure of model accuracy but correlation. Here the R^2 value is not low, but the RMSE is 3.1433×10^4 which means the average difference between model fitting and the observation is 3.1433×10^4. It may be a significant discrepancy from an application point of view. A high R^2 doesn't guarantee that the model has enough accuracy.

We used R^2 to show the impact of the error from independent and response variables in Chapter 7, where we didn't consider the impact of the number of parameters (because the number of parameters is very small compared to the number of observations). However, R^2 increases as the number of parameters increases. So people usually use adjusted R-squared, which is designed to mitigate the issue. The original R^2 is defined as:

$$R^2 = 1 - \frac{RSS}{TSS}$$

where $RSS = \sum_{i=1}^{n}(y_i - \hat{y}_i)^2$ and $TSS = \sum_{i=1}^{n}(y_i - \bar{y})^2$.

Since **Residual Sum of Squares (RSS)** is always decreasing as the number of parameters increases, R^2 increases as a result. For a model with p parameters, the adjusted R^2 is defined as:

$$Adjusted\ R^2 = 1 - \frac{RSS/(n-p-1)}{TSS/(n-1)}$$

To maximize the adjusted R^2 is identical to minimize $RSS/(n-p-1)$. Since the number of parameters p is reflected in the equation, $RSS/(n-p-1)$ can increase or decrease as p increases. The idea behind this is that the adjusted R-squared increases if the new variable improves the model more than would be expected by chance. It decreases when a predictor improves the model by less than expected by chance. While values are usually positive, they can be negative as well.

Another measurement is C_p. For a least squared model with p parameters:

$$C_p = \frac{1}{n}(RSS + 2p\hat{\sigma}^2)$$

where $\hat{\sigma}^2$ is the estimator of the model random effect ϵ. C_p is to add penalty $2p\hat{\sigma}^2$ to the training set RSS. The goal is to adjust the over-optimistic measurement based on training data. As the number of parameters increases, the penalty increases. It counteracts the decrease of RSS due to increasing the number of parameters. We choose the model with a smaller C_p.

Both **Akaike Information Criterion (AIC)** and **Bayesian Information Criterion (BIC)** are based on the maximum likelihood. In linear regression, the maximum likelihood estimate is the least squared estimate. The definitions of the two are

$$AIC = n + nlog(2\pi) + nlog(RSS/n) + 2(p+1)$$
$$BIC = n + nlog(2\pi) + nlog(RSS/n) + log(n)(p+1)$$

R function `AIC()` and `BIC()` will calculate the AIC and BIC value according to the above equations. Many textbooks ignore content item $n+nlog(2\pi)$, and use p instead of $p+1$. Those slightly different versions give the same results since we are only interested in the relative value. Comparing to AIC, BIC puts a heavier penalty on the number of parameters.

8.2 Classification Model Performance

This section focuses on performance measurement for models with a categorical response. The metrics in the previous section are for models with a continuous response and they are not appropriate in the context of classification. Most of the classification problems are dichotomous, such as an outbreak of disease, spam email, etc. There are also cases with more than two categories as the segments in the clothing company data. We use swine disease data to illustrate different metrics. Let's train a random forest model as an example. We will discuss the model in Chapter 11.

```
disease_dat <- read.csv("http://bit.ly/2KXb1Qi")
# you can check the data using glimpse()
# glimpse(disease_dat)
```

The process includes (1) separate the data to be training and testing sets, (2) fit model using training data (xTrain and yTrain), and (3) applied the trained model on testing data (xTest and yTest) to evaluate model performance.

We use 70% of the sample as training and the rest 30% as testing.

```
set.seed(100)
# separate the data to be training and testing
trainIndex <- createDataPartition(disease_dat$y, p = 0.8,
    list = F, times = 1)
xTrain <- disease_dat[trainIndex, ] %>% dplyr::select(-y)
xTest <- disease_dat[-trainIndex, ] %>% dplyr::select(-y)
# the response variable need to be factor
yTrain <- disease_dat$y[trainIndex] %>% as.factor()
yTest <- disease_dat$y[-trainIndex] %>% as.factor()
```

Train a random forest model:

```
train_rf <- randomForest(yTrain ~ .,
                          data = xTrain,
                          mtry = trunc(sqrt(ncol(xTrain) - 1)),
                          ntree = 1000,
                          importance = T)
```

Apply the trained random forest model to the testing data to get two types of predictions:

- probability (a value between 0 to 1)

```
yhatprob <- predict(train_rf, xTest, "prob")
set.seed(100)
car::some(yhatprob)
```

```
##              0     1
## 47    0.831 0.169
## 101   0.177 0.823
## 196   0.543 0.457
## 258   0.858 0.142
## 274   0.534 0.466
## 369   0.827 0.173
## 389   0.852 0.148
## 416   0.183 0.817
## 440   0.523 0.477
## 642   0.836 0.164
```

- category prediction (0 or 1)

```
yhat <- predict(train_rf, xTest)
car::some(yhat)
```

```
## 146 232 269 302 500 520 521 575 738 781
##   0   0   1   0   0   0   1   0   0   0
## Levels: 0 1
```

We will use the above two types of predictions to show different performance metrics.

8.2.1 Confusion Matrix

Confusion Matrix is a counting table to describe the performance of a classification model. For the true response yTest and prediction yhat, the confusion matrix is:

```
yhat = as.factor(yhat) %>% relevel("1")
yTest = as.factor(yTest) %>% relevel("1")
table(yhat, yTest)
```

```
##      yTest
## yhat  1  0
##    1 56  1
##    0 15 88
```

The top-left and bottom-right are the numbers of correctly classified samples. The top-right and bottom-left are the numbers of wrongly classified samples. A general confusion matrix for a binary classifier is following:

	Predicted Yes	Predicted No
Actual Yes	TP	FN
Actual No	FP	TN

where TP is true positive, FP is false positive, TN is true negative, FN is false negative. The cells along the diagonal line from top-left to bottom-right contain the counts of correctly classified samples. The cells along the other diagonal line contain the counts of wrongly classified samples. The most straightforward performance measure is the **total accuracy** which is the percentage of correctly classified samples:

$$Total\ accuracy = \frac{TP + TN}{TP + TN + FP + FN}$$

You can calculate the total accuracy when there are more than two categories. This statistic is straightforward but has some disadvantages. First, it doesn't differentiate different error types. In a real application, different types of error may have different impacts. For example, it is much worse to tag an important email as spam and miss it than failing to filter out a spam email. Provost et al. (1998) discussed in detail about the problem of using total accuracy on different classifiers. There are some other metrics based on the confusion matrix that measure different types of error.

Precision is a metric to measure how accurate positive predictions are (i.e. among those emails predicted as spam, how many percentages of them are spam emails?):

$$precision = \frac{TP}{TP + FP}$$

Sensitivity is to measure the coverage of actual positive samples (i.e. among those spam emails, how many percentages of them are predicted as spam) :

$$Sensitivity = \frac{TP}{TP + FN}$$

Specificity is to measure the coverage of actual negative samples (i.e. among those non-spam emails, how many percentages of them pass the filter):

$$Specificity = \frac{TN}{TN + FP}$$

Since wrongly tagging an important email as spam has a bigger impact, in the spam email case, we want to make sure the model specificity is high enough.

Second, total accuracy doesn't reflect the natural frequencies of each class. For example, the percentage of fraud cases for insurance may be very low, like 0.1%. A model can achieve nearly perfect accuracy (99.9%) by predicting all samples to be negative. The percentage of the largest class in the training set is also called the no-information rate. In this example, the no-information rate is 99.9%. You need to get a model that at least beats this rate.

8.2.2 Kappa Statistic

Another metric is the Kappa statistic. It measures the agreement between the observed and predicted classes. It was originally come up by Cohen etc. (Cohen, 1960). Kappa takes into account the accuracy generated simply by chance. It is defined as:

$$Kappa = \frac{P_0 - P_e}{1 - P_e}$$

Let $n = TP + TN + FP + FN$ be the total number of samples, where $P_0 = \frac{TP+TN}{n}$ is the observed accuracy, $P_e = \frac{(TP+FP)(TP+FN)+(FN+TN)(FP+TN)}{n^2}$ is the expected accuracy based on the marginal totals of the confusion matrix. Kappa can take on a value from -1 to 1. The higher the value, the higher the agreement. A value of 0 means there is no agreement between the observed and predicted classes, while a value of 1 indicates perfect agreement. A negative value indicates that the prediction is in the opposite direction of the observed value. The following table may help you "visualize" the interpretation of kappa (Landis and Koch, 1977):

Kappa	Agreement
<0	Less than chance agreement
0.01–0.20	Slight agreement
0.21–0.40	Fair agreement
0.41–0.60	Moderate agreement
0.61–0.80	Substantial agreement
0.81–0.99	Almost perfect agreement

In general, a value between 0.3 to 0.5 indicates a reasonable agreement. If a model has a high accuracy of 90%, while the expected accuracy is also high, say 85%. The Kappa statistics is $\frac{1}{3}$. It means the prediction and the observation have a fair

agreement. You can calculate Kappa when the number of categories
is larger than 2. The package `fmsb` has a function `Kappa.test()` to
calculate Cohen's Kappa statistics. The function can also return
the hypothesis test result and a confidence interval. Use the above
observation vector `yTest` and prediction vector `yhat` as an example,
you can calculate the statistics:

```
# install.packages("fmsb")
kt<-fmsb::Kappa.test(table(yhat,yTest))
kt$Result
```

```
##
##   Estimate Cohen's kappa statistics and test the
##   null hypothesis that the extent of agreement is
##   same as random (kappa=0)
##
## data:   table(yhat, yTest)
## Z = 9.7, p-value <2e-16
## 95 percent confidence interval:
##   0.6972 0.8894
## sample estimates:
## [1] 0.7933
```

The output of the above function contains an object named
`Judgement`:

```
kt$Judgement
```

```
## [1] "Substantial agreement"
```

8.2.3 ROC

Receiver Operating Characteristic (ROC) curve uses the predicted
class probabilities and determines an effective threshold such that
values above the threshold are indicative of a specific event. We

have shown the definitions of sensitivity and specificity above. The sensitivity is the true positive rate and specificity is true negative rate. "1 - specificity" is the false positive rate. ROC is a graph of pairs of true positive rate (sensitivity) and false positive rate (1-specificity) values that result as the test's cutoff value is varied. The Area Under the Curve (AUC) is a common measure for two-class problem. There is usually a trade-off between sensitivity and specificity. If the threshold is set lower, then there are more samples predicted as positive and hence the sensitivity is higher. Let's look at the predicted probability yhatprob in the swine disease example. The predicted probability object yhatprob has two columns, one is the predicted probability that a farm will have an outbreak, the other is the probability that farm will NOT have an outbreak. So the two add up to have value 1. We use the probability of outbreak (the 2nd column) for further illustration. You can use roc() function to get an ROC object (rocCurve) and then apply different functions on that object to get needed plot or ROC statistics. For example, the following code produces the ROC curve:

```
rocCurve <- pROC::roc(response = yTest,
              predictor = yhatprob[,2])


## Setting levels: control = 1, case = 0

## Setting direction: controls > cases



plot(1-rocCurve$specificities,
     rocCurve$sensitivities,
     type = 'l',
     xlab = '1 - Specificities',
     ylab = 'Sensitivities')
```

The first argument of the roc() is, response, the observation vector. The second argument is predictor is the continuous prediction

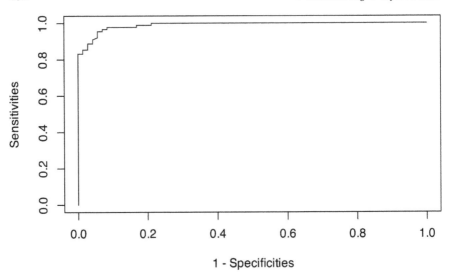

(probability or link function value). The x-axis of ROC curve is "1 - specificity" and the y-axis is "sensitivity." ROC curve starts from (0, 0) and ends with (1, 1). A perfect model that correctly identifies all the samples will have 100% sensitivity and specificity which corresponds to the curve that also goes through (0, 1). The area under the perfect curve is 1. A model that is totally useless corresponds to a curve that is close to the diagonal line and an area under the curve about 0.5.

You can visually compare different models by putting their ROC curves on one plot. Or use the AUC to compare them. DeLong et al. came up a statistic test to compare AUC based on U-statistics (E.R. DeLong, 1988) which can give a p-value and confidence interval. You can also use bootstrap to get a confidence interval for AUC (Hall et al., 2004).

We can use the following code in R to get an estimate of AUC and its confidence interval:

```
# get the estimate of AUC
auc(rocCurve)
```

```
## Area under the curve: 0.989
```

```
# get a confidence interval based on DeLong et al.
ci.auc(rocCurve)
```

```
## 95% CI: 0.979-0.999 (DeLong)
```

AUC is robust to class imbalance (Provost et al., 1998; Fawcett, 2006) hence a popular measurement. But it still boils a lot of information down to one number so there is inevitably a loss of information. It is better to double check by comparing the curve at the same time. If you care more about getting a model that will have high specificity, which is the lower part of the curve, as in the spam filtering case, you can use the area of the lower part of the curve as the performance measurement (McClish, 1989). ROC is only for two-class case. Some researchers generalized it to situations with more than two categories (Hand and Till, 2001; Lachiche and Flach, 2003; Li and Fine, 2008).

8.2.4 Gain and Lift Charts

Gain and lift chart is a visual tool for evaluating the performance of a classification model. In the previous swine disease example, there are 160 samples in the testing data and 89 of them have a positive outcome.

```
table(yTest)
```

```
## yTest
##  1  0
## 71 89
```

If we order the testing samples by the predicted probability, one would hope that the positive samples are ranked higher than the negative ones. That is what the lift charts do: rank the samples by their scores and calculate the cumulative positive rate as more samples are evaluated. In the perfect scenario, the highest-ranked 71 samples would contain all 71 positive samples. When the model is totally random, the highest-ranked x% of the data would contain about x% of the positive sample. The gain/lift charts compare the ratio between the results obtained with and without a model.

Let's plot the lift charts to compare the predicted outbreak probability (`modelscore <- yhatprob[ ,2]`) from random forest model we fit before with some random scores generated from a uniform distribution (`randomscore <- runif(length(yTest))`).

```
# predicted outbreak probability
modelscore <- yhatprob[ ,2]
# randomly sample from a uniform distribution
randomscore <- runif(length(yTest))
labs <- c(modelscore = "Random Forest Prediction",
          randomscore = "Random Number")
liftCurve <- caret::lift(yTest ~ modelscore + randomscore,
                 class = "1",
                 labels = labs)
xyplot(liftCurve, auto.key = list(columns = 2, lines = T, points = F))
```

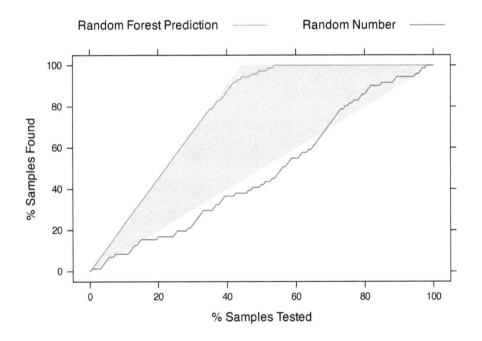

The x-axis is the percentage of samples tested and the y-axis is the percentage of positive samples that are detected by the model. For example, the point on the curve of random forest prediction, (8.125, 18.31), indicates that if you order the testing samples by the predicted probability from high to low, the top 8.125% of the samples contain 18.31% of the total positive outcomes.

Similar to the ROC curve, we can choose the model by comparing their lift charts. Some parts of the lift curves may be more interesting than the rest. For example, if you only have a budget to clean 50% of the farms, then you should pick the model that gives the highest point when the x-axis is 50%.

9

Regression Models

In this chapter, we will cover ordinary linear regression and a few more advanced regression methods. The linear combination of variables seems simple compared to many of today's machine learning models. However, many advanced models use linear combinations of variables as one of its major components or steps. For example, for each neuron in the deep neural network, all the input signals are first linearly combined before feeding to a non-linear activation function. To understand many of today's machine learning models, it is helpful to understand the key ideas across different modeling frameworks.

First, we will introduce multivariate linear regression (i.e. the typical least square regression) which is one of the simplest supervised learning methods. Even though it is simple, the general ideas and procedures of fitting a regression model are applied to a boarder scope. Having a solid understanding of the basic linear regression model enables us to learn more advanced models easily. For example, we will introduce two "shrinkage" versions of linear regression: ridge regression and LASSO regression. While the parameters are fitted by the least square method, the extra penalty can effectively shrink model parameters towards zero. It mediates overfitting and maintains the robustness of the model when data size is small compared to the number of explanatory variables. We first introduce basic knowledge of each model and then provide R codes to show how to fit the model. We only cover the major properties of these models and the listed reference will provide more in-depth discussion.

We will use the clothing company data as an example. We want to answer business questions such as "which variables are the driving factors of total revenue (both online and in-store purchase)?" The answer to this question can help the company to decide where

DOI: 10.1201/9781351132916-9

to invest (such as design, quality, etc). Note that the driving factor here does not guarantee a causal relationship. Linear regression models reveal correlation rather than causation. For example, if a survey on car purchase shows a positive correlation between price and customer satisfaction, does it suggest the car dealer should increase the price? Probably not! It is more likely that the customer satisfaction is impacted by quality. And a higher quality car tends to be more expensive. Causal inference is much more difficult to establish, and we have to be very careful when interpreting regression model results.

Load the R packages first:

```
# install packages from CRAN
p_needed <- c('caret', 'dplyr', 'lattice',
              'elasticnet', 'lars', 'corrplot',
              'pls')

packages <- rownames(installed.packages())
p_to_install <- p_needed[!(p_needed %in% packages)]
if (length(p_to_install) > 0) {
    install.packages(p_to_install)
}

lapply(p_needed, require, character.only = TRUE)
```

9.1 Ordinary Least Square

For a typical linear regression with p explanatory variables, we have a linear combinations of these variables:

$$f(\mathbf{X}) = \mathbf{X}\beta = \beta_0 + \sum_{j=1}^{p} \mathbf{x}_{.j}\beta_j$$

where β is the parameter vector with length $p+1$. Here we use $\mathbf{x}_{.j}$ for column vector and $\mathbf{x}_{i.}$ for row vector. Least square is the method to find a set of value for $\boldsymbol{\beta}^{\mathbf{T}} = (\beta_0, \beta_1, ..., \beta_p)$ such that it minimizes the residual sum of square (RSS):

$$RSS(\beta) = \sum_{i=1}^{N}(y_i - f(\mathbf{x_{i.}}))^2 = \sum_{i=1}^{N}(y_i - \beta_0 - \sum_{j=1}^{p} x_{ij}\beta_j)^2$$

The process of finding a set of values has been implemented in R. Now let's load the data:

```
dat <- read.csv("http://bit.ly/2P5gTw4")
```

Before fitting the model, we need to clean the data, such as removing bad data points that are not logical (negative expense).

```
dat <- subset(dat, store_exp > 0 & online_exp > 0)
```

Use 10 survey question variables as our explanatory variables.

```
modeldat <- dat[, grep("Q", names(dat))]
```

The response variable is the sum of in-store spending and online spending.

```
# total expense = in store expense + online expense
modeldat$total_exp <- dat$store_exp + dat$online_exp
```

To fit a linear regression model, let us first check if there are any missing values or outliers:

```
par(mfrow = c(1, 2))
hist(modeldat$total_exp, main = "", xlab = "total_exp")
boxplot(modeldat$total_exp)
```

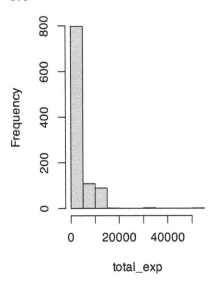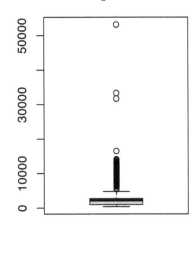

There is no missing value in the response variable, but there are outliers. Outliers are usually best described by the problem to solve itself such that we know from domain knowledge that it is not possible to have such values. We can also use a statistical threshold to remove extremely large or small outlier values from the data. We use the Z-score to find and remove outliers described in section 5.5. Readers can refer to the section for more detail.

```
y <- modeldat$total_exp
# Find data points with Z-score larger than 3.5
zs <- (y - mean(y))/mad(y)
modeldat <- modeldat[-which(zs > 3.5), ]
```

We will not perform log-transformation for the response variable at this stage. Let us first check the correlation among explanatory variables:

```
correlation <- cor(modeldat[, grep("Q", names(modeldat))])
corrplot::corrplot.mixed(correlation, order = "hclust", tl.pos = "lt",
    upper = "ellipse")
```

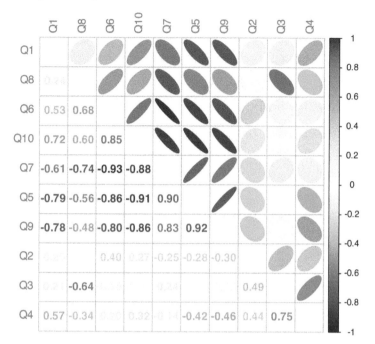

FIGURE 9.1
Correlation matrix plot for explanatory variables

As shown in figure 9.1, there are some highly correlated variables. Let us use the method described in section 5.6 to find potential highly correlated explanatory variables to remove with a threshold of 0.75:

```
highcor <- findCorrelation(correlation, cutoff = 0.75)
```

```
modeldat <- modeldat[, -highcor]
```

The dataset is now ready to fit a linear regression model. The standard format to define a regression in R is:

(1) response variable is at the left side of ~

(2) the explanatory variables are at the right side of ~

(3)   if all the variables in the dataset except the response variable are included in the model, we can use . at the right side of ~

(4)   if we want to consider the interaction between two variables such as Q1 and Q2, we can add an interaction term `Q1*Q2`

(5)   transformation of variables can be added directly to variable names such as `log(total_exp)`.

```
lmfit <- lm(log(total_exp) ~ ., data = modeldat)
summary(lmfit)
```

```
##
## Call:
## lm(formula = log(total_exp) ~ ., data = modeldat)
##
## Residuals:
##      Min       1Q   Median       3Q      Max
## -1.1749  -0.1372   0.0128   0.1416   0.5623
##
## Coefficients:
##              Estimate Std. Error t value Pr(>|t|)
## (Intercept)   8.09831    0.05429  149.18  < 2e-16 ***
## Q1           -0.14534    0.00882  -16.47  < 2e-16 ***
## Q2            0.10228    0.01949    5.25  2.0e-07 ***
## Q3            0.25445    0.01835   13.87  < 2e-16 ***
## Q6           -0.22768    0.01152  -19.76  < 2e-16 ***
## Q8           -0.09071    0.01650   -5.50  5.2e-08 ***
## ---
## Signif. codes:
## 0 '***' 0.001 '**' 0.01 '*' 0.05 '.' 0.1 ' ' 1
##
## Residual standard error: 0.226 on 805 degrees of freedom
## Multiple R-squared:  0.854,  Adjusted R-squared:  0.853
## F-statistic:  943 on 5 and 805 DF,  p-value: <2e-16
```

The summary(lmfit) presents a summary of the model fit. It shows the point estimate of each explanatory variable (the Estimate column), their corresponding standard error (the Std. Error column), t values (t value), and p values (Pr(>|t|)).

9.1.1 The Magic P-value

Let us pause a little to have a short discussion about p-value. Misuse of p-value is common in many research fields. There were heated discussions about P-value in the past. Siegfried commented in his 2010 Science News article:

"It's science's dirtiest secret: The scientific method of testing hypotheses by statistical analysis stands on a flimsy foundation."

American Statistical Association (i.e., ASA) released an official statement on p-value in 2016 (R. L. Wassersteina, 2016). It was the first time to have an organization level announcement about p-value. ASA stated that the goal to release this guidance was to

---

"improve the conduct and interpretation of quantitative science and inform the growing emphasis on reproducibility of science research."

The statement also noted that

"the increased quantification of scientific research and a proliferation of large, complex data sets has expanded the scope for statistics and the importance of appropriately chosen techniques, properly conducted analyses, and correct interpretation."

The statement's six principles, many of which address misconceptions and misuse of the P-value, are the following:

1. P-values can indicate how incompatible the data are with a specified statistical model.
2. P-values do not measure the probability that the studied hypothesis is true or the probability that the data were produced by random chance alone.
3. Scientific conclusions and business or policy decisions should not be based only on whether a p-value passes a specific threshold.
4. Proper inference requires full reporting and transparency.
5. A p-value, or statistical significance, does not measure the size of an effect or the importance of a result.
6. By itself, a p-value does not provide a good measure of evidence regarding a model or hypothesis.

The $p = 0.05$ threshold is not based on any scientific calculation but is an arbitrary number. It means that practitioners can use a different threshold if they think it better fits the problem to solve. We do not promote p-value in this book. However, the p-value is hard to avoid in classical statistical inference. In practice, when making classic statistical inferences, we recommend reporting confidence interval whenever possible instead of P-value.

The Bayesian paradigm is an alternative to the classical paradigm. A Bayesian can state probabilities about the parameters, which are considered random variables. However, it is not possible in the classical paradigm. In our work, we use hierarchical (generalize) linear models in practice instead of classical linear regression. Hierarchical models pool information across clusters (for example, you can treat each customer segment as a cluster). This pooling tends to improve estimates of each cluster, especially when sampling is imbalanced. Because the models automatically cope with differing uncertainty introduced by sampling imbalance (bigger cluster has smaller variance), it prevents over-sampled clusters from unfairly dominating inference.

This book does not cover the Bayesian framework. The best applied Bayesian book is Statistical Rethinking[1] by Richard McElreath (McElreath, 2020). The book provides outstanding conceptual explanations and a wide range of models from simple to advanced with detailed, repeatable code. The text uses R, but there are code examples for Python and Julia on the book website.

Now let us come back to our example. We will not spend too much time on p-values, while we will focus on the confidence interval for the parameter estimate for each explanatory variable. In R, the function confint() can produce the confidence interval for each parameter:

```
confint(lmfit,level=0.9)
```

```
##                   5 %      95 %
## (Intercept)   8.00892   8.18771
## Q1           -0.15987  -0.13081
## Q2            0.07018   0.13437
## Q3            0.22424   0.28466
## Q6           -0.24665  -0.20871
## Q8           -0.11787  -0.06354
```

The above output is for a 90% confidence level as level=0.9 indicated in the function call. We can change the confidence level by adjusting the level setting.

Fitting a linear regression is so easy using R that many analysts directly write reports without thinking about whether the model is meaningful. On the other hand, we can easily use R to check model assumptions. In the following sections, we will introduce a few commonly used diagnostic methods for linear regression to check whether the model assumptions are reasonable.

9.1.2 Diagnostics for Linear Regression

In linear regression, we would like the Ordinary Least Square (OLS) estimate to be the Best Linear Unbiased Estimate (BLUE). In other

---

[1] https://xcelab.net/rm/statistical-rethinking/

words, we hope the expected value of the estimate is the actual parameter value (i.e., unbiased) and achieving minimized residual (i.e., best). Based on the Gauss-Markov theorem, the OLS estimate is BLUE under the following conditions:

1. Explanatory variables $(\mathbf{x}_{.j})$ and random error (ϵ) are independent: $cov(\mathbf{x}_{.j}, \epsilon) = 0$ for $\forall j = j \in 1...p$.

2. The expected value of random error is zero: $E(\epsilon|\mathbf{X}) = 0$

3. Random errors are uncorrelated with each other, and the variance of random error is consistent: $Var(\epsilon) = \sigma^2 I$, where σ is positive and I is a $n \times n$ identical matrix.

We will introduce four graphic diagnostics for the above assumptions.

(1)   Residual plot

It is a scatter plot with residual on the Y-axis and fitted value on the X-axis. We can also put any of the explanatory variables on the X-axis. Under the assumption, residuals are randomly distributed, and we need to check the following:

- Are residuals centered around zero?
- Are there any patterns in the residual plots (such as residuals with x-values farther from $\bar{x}$ have greater variance than residuals with x-values closer to $\bar{x}$)?
- Are the variances of the residual consistent across a range of fitted values?

Please note that even if the variance is not consistent, the regression parameter's point estimate is still unbiased. However, the variance estimate is not unbiased. Because the significant test for regression parameters is based on the random error distribution, these tests are no longer valid if the variance is not constant.

(2)   Normal quantile-quantile Plot (Q-Q Plot)

Q-Q Plot is used to check the normality assumption for the residual. For normally distributed residuals, the data points should follow a straight line along the Q-Q plot. The more departure from a straight line, the more departure from a normal distribution for the residual.

(3) Standardized residuals plot

Standardized residual is the residual normalized by an estimate of its standard deviation. Like the residual plot, the X-axis is still the fitted value, but the y-axis is now standardized residuals. Because of the normalization, the y-axis shows the number of standard deviations from zero. A value greater than 2 or less than -2 indicates observations with large standardized residuals. The plot is useful because when the variance is not consistent, it can be difficult to detect the outliers using the raw residuals plot.

(4) Cook's distance

Cook's distance can check influential points in OLS based linear regression models. In general, we need to pay attention to data points with Cook's distance > 0.5.

In R, these diagnostic graphs are built in the `plot()` function.

```
par(mfrow = c(2, 2))
plot(lmfit, which = 1)
plot(lmfit, which = 2)
plot(lmfit, which = 3)
plot(lmfit, which = 4)
```

The above diagnostic plot examples show:

- Residual plot: residuals are generally distributed around $y = 0$ horizontal line. There are no significant trends or patterns in this residual plot (there are two bumps but does not seem too severe). So the linear relationship assumption between the response variable and explanatory variables is reasonable.

- Q-Q plot: data points are pretty much along the diagonal line of $Y = X$, indicating no significant normality assumption departure for the residuals. Because we simulate the data, we know the response variable before log transformation follows a normal distribution. The shape of the distribution does not deviate from a normal distribution too much after log transformation.

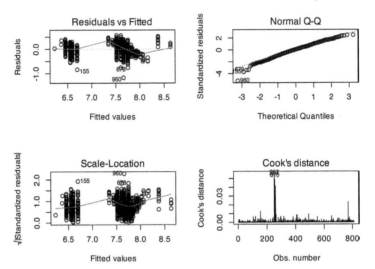

FIGURE 9.2
Linear regression diagnostic plots: residual plot (top left), Q-Q plot
(top right), standardized residuals plot (lower left), Cook's distance
(lower right)

Note that the Gauss-Markov theorem does not require normality.
We need the normal assumption to look for significant factors or
define a confidence interval. However, as Andrew Gelman pointed
out in section 3.6 of his book (Gelman and Hill, 2006), normality
and equal variance are typically minor concerns.

- Standardized residual plot: if the constant variance assumption is valid, then the plot's data points should be randomly
  distributed around the horizontal line. We can see there are
  three outliers on the plot. Let us check those points:

```
modeldat[which(row.names(modeldat) %in% c(960, 678, 155)), ]
```

```
##      Q1 Q2 Q3 Q6 Q8 total_exp
## 155   4  2  1  4  4     351.9
## 678   2  1  1  1  2    1087.3
## 960   2  1  1  1  3     658.3
```

It is not easy to see why those records are outliers from the above output. It will be clear conditional on the independent variables (Q1, Q2, Q3, Q6, and Q8). Let us examine the value of total_exp for samples with the same Q1, Q2, Q3, Q6, and Q8 answers as the 3rd row above.

```
datcheck = modeldat %>%
    filter(Q1 ==2 & Q2 == 1 & Q3 == 1 & Q6 == 1 & Q8 == 3)
nrow(datcheck)
```

```
## [1] 87
```

There are 87 samples with the same values of independent variables. The response variable's (total_exp) distribution is:

```
summary(datcheck$total_exp)
```

```
##     Min. 1st Qu.  Median    Mean 3rd Qu.    Max.
##      658    1884    2215    2204    2554    3197
```

Now it is easy to see why row 960 with total_exp = 658.3 is an outlier. All the other 86 records with the same survey responses have a much higher total expense!

- Cook's distance: the maximum of Cook's distance is around 0.05. Even though the graph does not have any point with Cook's distance of more than 0.5, we could spot some outliers.

The graphs suggest some outliers, but it is our decision what to do. We can either remove it or investigate it further. If the values are not due to any data error, we should consider them in our analysis.

9.2 Principal Component Regression and Partial Least Square

In real-life applications, explanatory variables are usually related to each other, containing similar information. For example, in the previous chapter, we used expenditure variables to predict consumer income. In that model, store expenditure (store_exp), online expenditure (online_exp), number of store transactions (store_trans), and number of online transactions (online_trans) are correlated to a certain extent, especially the number of transactions and expenditure. If there is a strong correlation among explanatory variables, then the least square-based linear regression model may not be robust. If the number of observations is less than the number of explanatory variables, the standard least square method cannot provide a unique set of coefficient estimates. We can perform data preprocessing, such as remove highly correlated variables with a preset threshold for such cases. However, this approach cannot guarantee a low correlation of the linear combination of the variables with other variables. In that case, the standard least square method will still not be robust. We need to be aware that removing highly correlated variables is not always guarantee a robust solution. We can also apply feature engineering procedures to explanatory variables such as principal component analysis (PCA). By using principal components, we can ensure they are uncorrelated with each other. However, the drawback of using PCA is that these components are linear combinations of original variables, and it is difficult to explain the fitted model. Principal component regression (PCR) is described in more detail in (Massy, 1965). It can be used when there are strong correlations among variables or when the number of observations is less than the number of variables.

In theory, we can use PCR to reduce the number of variables used in a linear model, but the results are not good. Because the first a few principal components may not be good candidates to model the response variable. PCA is unsupervised learning such that the entire process does not consider the response variable. In PCA, it only focuses on the variability of explanatory variables.

When the independent variables and response variables are related, PCR can well identify the systematic relationship between them. However, when there exist independent variables not associated with response variable, it will undermine PCR's performance. We need to be aware that PCR does not make feature selections, and each principal component is a combination of original variables.

Partial least square regression (PLS) is the supervised version of PCR. Similar to PCR, PLS can also reduce the number of variables in the regression model. As PLS is also related to the variables' variance, we usually standardize or normalize variables before PLS. Suppose we have a list of explanatory variables $\mathbf{X} = [X_1, X_2, ..., X_p]^T$, and their variance-covariance matrix is Σ. PLS also transforms the original variables using linear combination to new uncorrelated variables $(Z_1, Z_2, ..., Z_m)$. When $m = p$, the result of PLS is the same as OLS. The main difference between PCR and PLS is the process of creating new variables. PLS considers the response variable.

PLS is from Herman Wold's Nonlinear Iterative Partial Least Squares (NIPALS) algorithm (Wold, 1973; Wold and Jöreskog, 1982). Later NIPALS was applied to regression problems, which was then called PLS. PLS is a method of linearizing nonlinear relationships through hidden layers. It is similar to the PCR, except that PCR does not take into account the information of the dependent variable when selecting the components. PCR's purpose is to find the linear combinations (i.e., unsupervised) that capture the most variance of the independent variables, while PLS maximizes the linear combination of dependencies with the response variable. In the current case, the more complicated PLS does not perform better than simple linear regression. We will not discuss the PLS algorithm's detail, and the reference mentioned above provides a more detailed description of the algorithm.

We focus on using R library `caret` to fit PCR and PLS models. Let us use the 10 survey questions (`Q1-Q10`) as the explanatory variables and income as the response variable. First load the data and preprocessing the data:

```r
library(lattice)
library(caret)
library(dplyr)
library(elasticnet)
library(lars)

# Load Data
sim.dat <- read.csv("http://bit.ly/2P5gTw4")
ymad <- mad(na.omit(sim.dat$income))

# Calculate Z values
zs <- (sim.dat$income - mean(na.omit(sim.dat$income)))/ymad
# which(na.omit(zs>3.5)) find outlier
# which(is.na(zs)) find missing values
idex <- c(which(na.omit(zs > 3.5)), which(is.na(zs)))
# Remove rows with outlier and missing values
sim.dat <- sim.dat[-idex, ]
```

Now let's define explanatory variable matrix (xtrain) by selecting these 10 survey questions columns, and define response variable (ytrain):

```r
xtrain = dplyr::select(sim.dat, Q1:Q10)
ytrain = sim.dat$income
```

We also set up random seed and 10-folder cross-validation:

```r
set.seed(100)
ctrl <- trainControl(method = "cv", number = 10)
```

Fit PLS model using number of explanatory variables as the hyper-parameter to tune. As there are at most 10 explanatory variables in the model, we set up the hyper-parameter tuning range to be 1 to 10:

```
plsTune <- train(xtrain, ytrain,
                 method = "pls",
                 # set hyper-parameter tuning range
                 tuneGrid = expand.grid(.ncomp = 1:10),
                 trControl = ctrl)
plsTune
```

```
## Partial Least Squares
##
## 772 samples
##  10 predictor
##
## No pre-processing
## Resampling: Cross-Validated (10 fold)
## Summary of sample sizes: 696, 693, 694, 694, 696, 695, ...
## Resampling results across tuning parameters:
##
##   ncomp  RMSE    Rsquared  MAE
##   1      27777   0.6534    19845
##   2      24420   0.7320    15976
##   3      23175   0.7590    14384
##   4      23011   0.7625    13808
##   5      22977   0.7631    13737
##   6      22978   0.7631    13729
##   7      22976   0.7631    13726
##   8      22976   0.7631    13726
##   9      22976   0.7631    13726
##   10     22976   0.7631    13726
##
## RMSE was used to select the optimal model using
##   the smallest value.
## The final value used for the model was ncomp = 7.
```

From the result, we can see that the optimal number of variables is 7. However, if we pay attention to the RMSE improvement, we will find only minimum improvement in RMSE after three variables.

In practice, we could choose to use the model with three variables if the improvement does not make a practical difference, and we would rather have a simpler model.

We can also find the relative importance of each variable during PLS model tuning process, as described using the following code:

```
plsImp <- varImp(plsTune, scale = FALSE)
plot(plsImp, top = 10, scales = list(y = list(cex = 0.95)))
```

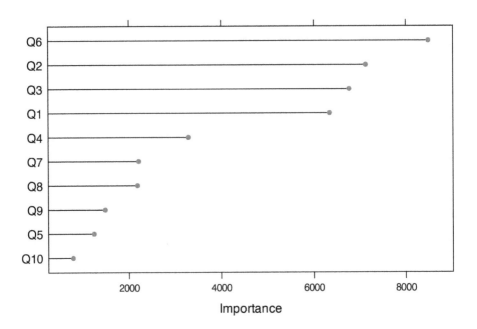

The above plot shows that Q1, Q2, Q3, and Q6, are more important than other variables. Now let's fit a PCR model with number of principal components as the hyper-parameter:

```
# Set random seed
 set.seed(100)
 pcrTune <- train(x = xtrain, y = ytrain,
         method = "pcr",
          # set hyper-parameter tuning range
```

```
            tuneGrid = expand.grid(.ncomp = 1:10),
            trControl = ctrl)
  pcrTune
```

```
## Principal Component Analysis
##
## 772 samples
##  10 predictor
##
## No pre-processing
## Resampling: Cross-Validated (10 fold)
## Summary of sample sizes: 696, 693, 694, 694, 696, 695, ...
## Resampling results across tuning parameters:
##
##   ncomp  RMSE   Rsquared  MAE
##    1     45958  0.03243   36599
##    2     32460  0.52200   24041
##    3     23235  0.75774   14516
##    4     23262  0.75735   14545
##    5     23152  0.75957   14232
##    6     23133  0.76004   14130
##    7     23114  0.76049   14129
##    8     23115  0.76045   14130
##    9     22991  0.76283   13801
##   10     22976  0.76308   13726
##
## RMSE was used to select the optimal model using
##   the smallest value.
## The final value used for the model was ncomp = 10.
```

From the output, the default recommendation is ten components. However, if we pay attention to RMSE improvement with more components, we will find little difference after the model with three components. Again, in practice, we can keep models with three components.

Now let's compare the hyper-parameter tuning process for PLS and PCR:

```
# Save PLS model tuning information to plsResamples
plsResamples <- plsTune$results
plsResamples$Model <- "PLS"
# Save PCR model tuning information to plsResamples
pcrResamples <- pcrTune$results
pcrResamples$Model <- "PCR"
# Combine both output for plotting
plsPlotData <- rbind(plsResamples, pcrResamples)
# Leverage xyplot() function from lattice library
xyplot(RMSE ~ ncomp,
       data = plsPlotData,
       xlab = "# Components",
       ylab = "RMSE (Cross-Validation)",
       auto.key = list(columns = 2),
       groups = Model,
       type = c("o", "g"))
```

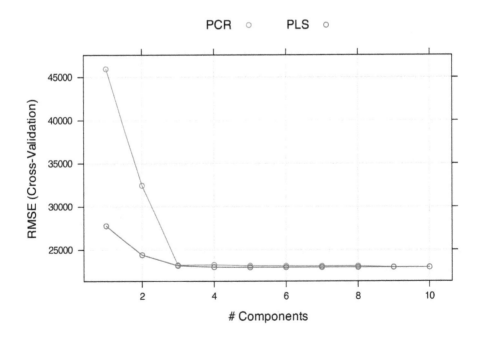

The plot confirms our choice of using a model with three components for both PLS and PCR.

10

Regularization Methods

The regularization method is also known as the shrinkage method. It is a technique that constrains or regularizes the coefficient estimates. By imposing a penalty on the size of coefficients, it shrinks the coefficient estimates towards zero. It also intrinsically conduct feature selection and is naturally resistant to non-informative predictors. It may not be obvious why this technique improves model performance, but it turns out to be a very effective modeling technique. In this chapter, we will introduce two best-known regularization methods: ridge regression and lasso. The elastic net is a combination of ridge and lasso, or it is a general representation of the two.

We talked about the variance bias trade-off in section 7.1. The variance of a learning model is the amount by which $\hat{f}$ would change if we estimated it using a different training data set. In general, model variance increases as flexibility increases. The regularization technique decreases the model flexibility by shrinking the coefficient and hence significantly reduce the model variance.

Load the R packages:

```
# install packages from CRAN
p_needed <- c('caret', 'elasticnet', 'glmnet', 'devtools',
              'MASS', 'grplasso')

packages <- rownames(installed.packages())
p_to_install <- p_needed[!(p_needed %in% packages)]
if (length(p_to_install) > 0) {
    install.packages(p_to_install)
}
```

DOI: 10.1201/9781351132916-10

```
lapply(p_needed, require, character.only = TRUE)

# install packages from GitHub
p_needed_gh <- c('NetlifyDS')

if (! p_needed_gh %in% packages) {
    devtools::install_github("netlify/NetlifyDS")
}

library(NetlifyDS)
```

10.1   Ridge Regression

Recall that the least square estimates minimize RSS:

$$RSS = \Sigma_{i=1}^{n}(y_i - \beta_0 - \Sigma_{j=1}^{p}\beta_j x_{ij})^2$$

Ridge regression (Hoerl and Kennard, 1970) is similar but it finds $\hat{\beta}^R$ that optimizes a slightly different function:

$$\Sigma_{i=1}^{n}(y_i - \beta_0 - \Sigma_{j=1}^{p}\beta_j x_{ij})^2 + \lambda\Sigma_{j=1}^{p}\beta_j^2 = RSS + \lambda\Sigma_{j=1}^{p}\beta_j^2$$

$$(10.1)$$

where $\lambda > 0$ is a tuning parameter. As with the least squares, ridge regression considers minimizing RSS. However, it adds a shrinkage penalty $\lambda\Sigma_{j=1}^{p}\beta_j^2$ that takes account of the number of parameters in the model. When $\lambda = 0$, it is identical to least squares. As λ gets larger, the coefficients start shrinking towards 0. When $\lambda \to \infty$, the rest of the coefficients $\beta_1, ..., \beta_p$ are close to 0. Here, the penalty is not applied to β_0. The tuning parameter λ is used to adjust the impact of the two parts in equation (10.1). Every value of λ corresponds to a set of parameter estimates.

There are many R packages for ridge regression, such as `lm.ridge()` function from MASS, function `enet()` from elasticnet. If you know the value of λ, you can use either of the function to fit ridge regression. A more convenient way is to use `train()` function from caret. Let's use the 10 survey questions to predict the total purchase amount (sum of online and store purchase).

```r
dat <- read.csv("http://bit.ly/2P5gTw4")
# data cleaning: delete wrong observations
# expense can't be negative
dat <- subset(dat, store_exp > 0 & online_exp > 0)
# get predictors
trainx <- dat[ , grep("Q", names(dat))]
# get response
trainy <- dat$store_exp + dat$online_exp
```

Use `train()` function to tune parameter. Since ridge regression adds the penalty parameter λ in front of the sum of squares of the parameters, the scale of the parameters matters. So here it is better to center and scale the predictors. This preprocessing is generally recommended for all techniques that puts penalty to parameter estimates. In this example, the 10 survey questions are already with the same scale so data preprocessing doesn't make too much different. It is a good idea to set the preprocessing as a standard.

```r
# set cross validation
ctrl <- trainControl(method = "cv", number = 10)
# set the parameter range
ridgeGrid <- data.frame(.lambda = seq(0, .1, length = 20))
set.seed(100)
ridgeRegTune <- train(trainx, trainy,
                      method = "ridge",
                      tuneGrid = ridgeGrid,
                      trControl = ctrl,
                      ## center and scale predictors
                      preProc = c("center", "scale"))
ridgeRegTune
```

```
## Ridge Regression
##
## 999 samples
##  10 predictor
##
## Pre-processing: centered (10), scaled (10)
## Resampling: Cross-Validated (10 fold)
## Summary of sample sizes: 899, 899, 899, 899, 899, 900, ...
## Resampling results across tuning parameters:
##
##   lambda    RMSE  Rsquared  MAE
##   0.000000  1744  0.7952    754.0
##   0.005263  1744  0.7954    754.9
##   0.010526  1744  0.7955    755.9
##   0.015789  1744  0.7955    757.3
##   0.021053  1745  0.7956    758.8
##   0.026316  1746  0.7956    760.6
##   0.031579  1747  0.7956    762.4
##   0.036842  1748  0.7956    764.3
##   0.042105  1750  0.7956    766.4
##   0.047368  1751  0.7956    768.5
##   0.052632  1753  0.7956    770.6
##   0.057895  1755  0.7956    772.7
##   0.063158  1757  0.7956    774.9
##   0.068421  1759  0.7956    777.2
##   0.073684  1762  0.7956    779.6
##   0.078947  1764  0.7955    782.1
##   0.084211  1767  0.7955    784.8
##   0.089474  1769  0.7955    787.6
##   0.094737  1772  0.7955    790.4
##   0.100000  1775  0.7954    793.3
##
## RMSE was used to select the optimal model using
##  the smallest value.
## The final value used for the model was lambda
##  = 0.005263.
```

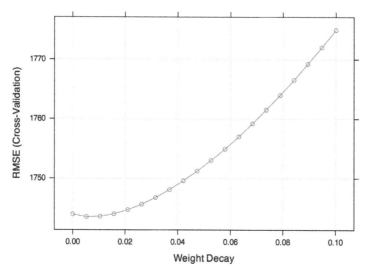

FIGURE 10.1
Test mean squared error for the ridge regression

The results show that the best value of λ is 0.005 and the RMSE and R^2 are 1744 and 0.7954 correspondingly. You can see from the figure 10.1, as the λ increase, the RMSE first slightly decreases and then increases.

```
plot(ridgeRegTune)
```

Once you have the tuning parameter value, there are different functions to fit a ridge regression. Let's look at how to use enet() in elasticnet package.

```
ridgefit = enet(x = as.matrix(trainx), y = trainy, lambda = 0.01,
                # center and scale predictors
                normalize = TRUE)
```

Note here ridgefit only assigns the value of the tuning parameter for ridge regression. Since the elastic net model include both ridge and lasso penalty, we need to use predict() function to get the model fit. You can get the fitted results by setting s = 1 and mode = "fraction". Here s = 1 means we only use the ridge parameter. We will come back to this when we get to lasso regression.

```
ridgePred <- predict(ridgefit, newx = as.matrix(trainx),
                     s = 1, mode = "fraction", type = "fit")
```

By setting type = "fit", the above returns a list object. The fit item has the predictions:

```
names(ridgePred)
```

```
## [1] "s"        "fraction" "mode"      "fit"
```

```
head(ridgePred$fit)
```

```
##       1      2      3      4      5      6
## 1290.5  224.2  591.4 1220.6  853.4  908.2
```

If you want to check the estimated coefficients, you can set type="coefficients":

```
ridgeCoef<-predict(ridgefit,newx = as.matrix(trainx),
                   s=1, mode="fraction", type="coefficients")
```

It also returns a list and the estimates are in the coefficients item:

```
# didn't show the results
RidgeCoef = ridgeCoef$coefficients
```

Comparing to the least square regression, ridge regression performs better because of the bias-variance-trade-off we mentioned in section 7.1. As the penalty parameter λ increases, the flexibility of the ridge regression decreases. It decreases the variance of the model but increases the bias at the same time.

10.2 LASSO

Even though the ridge regression shrinks the parameter estimates towards 0, it won't shink any estimates to be exactly 0 which means it includes all predictors in the final model. So it can't select variables. It may not be a problem for prediction but it is a huge disadvantage if you want to interpret the model especially when the number of variables is large. A popular alternative to the ridge penalty is the **Least Absolute Shrinkage and Selection Operator** (LASSO) (Tibshirani, 1996).

Similar to ridge regression, lasso adds a penalty. The lasso coefficients $\hat{\beta}_{\lambda}^{L}$ minimize the following:

$$\Sigma_{i=1}^{n}(y_i - \beta_0 - \Sigma_{j=1}^{p}\beta_j x_{ij})^2 + \lambda\Sigma_{j=1}^{p}|\beta_j| = RSS + \lambda\Sigma_{j=1}^{p}|\beta_j| \tag{10.2}$$

The only difference between lasso and ridge is the penalty. In statistical parlance, ridge uses L_2 penalty (β_j^2) and lasso uses L_1 penalty ($|\beta_j|$). L_1 penalty can shrink the estimates to 0 when λ is big enough. So lasso can be used as a feature selection tool. It is a huge advantage because it leads to a more explainable model.

Similar to other models with tuning parameters, lasso regression requires cross-validation to tune the parameter. You can use `train()` in a similar way as we showed in the ridge regression section. To tune parameter, we need to set cross-validation and parameter range. Also, it is advised to standardize the predictors:

```
ctrl <- trainControl(method = "cv", number = 10)
lassoGrid <- data.frame(fraction = seq(.8, 1, length = 20))
set.seed(100)
lassoTune <- train(trainx, trainy,
                   ## set the method to be lasso
                   method = "lars",
                   tuneGrid = lassoGrid,
                   trControl = ctrl,
```

```
                              ## standardize the predictors
                              preProc = c("center", "scale"))
lassoTune
```

```
## Least Angle Regression
##
## 999 samples
##  10 predictor
##
## Pre-processing: centered (10), scaled (10)
## Resampling: Cross-Validated (10 fold)
## Summary of sample sizes: 899, 899, 899, 899, 899, 900, ...
## Resampling results across tuning parameters:
##
##    fraction  RMSE  Rsquared  MAE
##    0.8000    1763  0.7921    787.5
##    0.8105    1760  0.7924    784.1
##    0.8211    1758  0.7927    780.8
##    0.8316    1756  0.7930    777.7
##    0.8421    1754  0.7933    774.6
##    0.8526    1753  0.7936    771.8
##    0.8632    1751  0.7939    769.1
##    0.8737    1749  0.7942    766.6
##    0.8842    1748  0.7944    764.3
##    0.8947    1746  0.7947    762.2
##    0.9053    1745  0.7949    760.1
##    0.9158    1744  0.7951    758.3
##    0.9263    1743  0.7952    756.7
##    0.9368    1743  0.7953    755.5
##    0.9474    1742  0.7954    754.5
##    0.9579    1742  0.7954    754.0
##    0.9684    1742  0.7954    753.6
##    0.9789    1743  0.7953    753.4
##    0.9895    1743  0.7953    753.5
##    1.0000    1744  0.7952    754.0
##
```

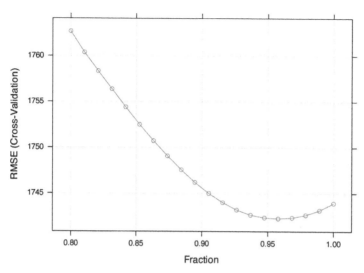

FIGURE 10.2
Test mean squared error for the lasso regression

```
## RMSE was used to select the optimal model using
##   the smallest value.
## The final value used for the model was fraction
##   = 0.9579.
```

The results show that the best value of the tuning parameter
(fraction from the output) is 0.957 and the RMSE and R^2 are 1742
and 0.7954 correspondingly. The performance is nearly the same
with ridge regression. You can see from the figure 10.2, as the λ
increase, the RMSE first decreases and then increases.

```
plot(lassoTune)
```

Once you select a value for tuning parameter, there are different
functions to fit lasso regression, such as lars() in lars, enet() in
elasticnet, glmnet() in glmnet. They all have very similar syntax.
Here we continue using enet(). The syntax is similar to ridge
regression. The only difference is that you need to set lambda = 0
because the argument lambda here is to control the ridge penalty.
When it is 0, the function will return the lasso model object.

```
lassoModel <- enet(x = as.matrix(trainx), y = trainy,
                   lambda = 0, normalize = TRUE)
```

Set the fraction value to be 0.957 (the value we got above):

```
lassoFit <- predict(lassoModel, newx = as.matrix(trainx),
                    s = 0.957, mode = "fraction", type = "fit")
```

Again by setting `type = "fit"`, the above returns a list object. The `fit` item has the predictions:

```
head(lassoFit$fit)
```

```
##      1      2      3      4      5      6
## 1357.3  300.5  690.2 1228.2  838.4 1010.1
```

You need to set `type = "coefficients"` to get parameter estimates:

```
lassoCoef <- predict(lassoModel,
                     newx = as.matrix(trainx),
                     s = 0.95,
                     mode = "fraction",
                     type = "coefficients")
```

It also returns a list and the estimates are in the `coefficients` item:

```
# didn't show the results
LassoCoef = lassoCoef$coefficients
```

Many researchers applied lasso to other learning methods, such as linear discriminant analysis (Line et al., 2011), partial least squares regression(Chun and Keleş, 2010). However, since the L_1 norm is not differentiable, optimization for lasso regression is more complicated. People come up with different algorithms to solve the computation problem. The biggest breakthrough is Least Angle Regression (LARS) from Bradley Efron etc. This algorithm works well for lasso regression especially when the dimension is high.

10.3   Elastic Net

Elastic Net is a generalization of lasso and ridge regression (Zou and Hastie, 2005). It combines the two penalties. The estimates of coefficients optimize the following function:

$$\Sigma_{i=1}^{n}(y_i - \hat{y}_i)^2 + \lambda_1\Sigma_{j=1}^{p}\beta_j^2 + \lambda_2\Sigma_{j=1}^{p}|\beta_j| \qquad (10.3)$$

Ridge penalty shrinks the coefficients of correlated predictors towards each other while the lasso tends to pick one and discard the others. So lasso estimates have a higher variance. However, ridge regression doesn't have a variable selection property. The advantage of the elastic net is that it keeps the feature selection quality from the lasso penalty as well as the effectiveness of the ridge penalty. And it deals with highly correlated variables more effectively.

We can still use `train()` function to tune the parameters in the elastic net. As before, set the cross-validation and parameter range, and standardize the predictors:

```
enetGrid <- expand.grid(.lambda = seq(0,0.2,length=20),
                        .fraction = seq(.8, 1, length = 20))
set.seed(100)
enetTune <- train(trainx, trainy,
                  method = "enet",
                  tuneGrid = enetGrid,
```

```
                       trControl = ctrl,
                       preProc = c("center", "scale"))
enetTune
```

Elasticnet

999 samples
 10 predictor

Pre-processing: centered (10), scaled (10)
Resampling: Cross-Validated (10 fold)
Summary of sample sizes: 899, 899, 899, 899, 899, 900, ...
Resampling results across tuning parameters:

lambda	fraction	RMSE	Rsquared	MAE
0.00000	0.8000	1763	0.7921	787.5
0.00000	0.8105	1760	0.7924	784.1
.				
.				
.				
0.09474	0.9158	1760	0.7945	782.5
0.09474	0.9263	1761	0.7947	782.5
0.09474	0.9368	1761	0.7949	782.7
0.09474	0.9474	1763	0.7950	783.3
0.09474	0.9579	1764	0.7951	784.3
0.09474	0.9684	1766	0.7953	785.7
0.09474	0.9789	1768	0.7954	787.1
0.09474	0.9895	1770	0.7954	788.8
0.09474	1.0000	1772	0.7955	790.4

[reached getOption("max.print") -- omitted 200 rows]

RMSE was used to select the optimal model using the smallest value.
The final values used for the model were fraction = 0.9579 and lambda = 0.

The results show that the best values of the tuning parameters are fraction $= 0.9579$ and lambda $= 0$. It also indicates that the final model is lasso only (the ridge penalty parameter lambda is 0). The RMSE and R^2 are 1742.2843 and 0.7954 correspondingly.

10.4 Penalized Generalized Linear Model

Adding penalties is a general technique that can be applied to many methods other than linear regression. In this section, we will introduce the penalized generalized linear model. It is to fit the generalized linear model by minimizing a penalized maximum likelihood. The penalty can be L_1, L_2 or a combination of the two. The estimates of coefficients minimize the following:

$$\min_{\beta_0, \beta} \frac{1}{N} \Sigma_{i=1}^{N} w_i l(y_i, \beta_0 + \boldsymbol{\beta}^{\mathbf{T}} \mathbf{x_i}) + \lambda[(1 - \alpha) \parallel \boldsymbol{\beta} \parallel_2^2 / 2 + \alpha \parallel \boldsymbol{\beta} \parallel_1]$$

where

$$l(y_i, \beta_0 + \boldsymbol{\beta}^{\mathbf{T}} \mathbf{x_i}) = -log[\mathcal{L}(y_i, \beta_0 + \boldsymbol{\beta}^{\mathbf{T}} \mathbf{x_i})]$$

It is the negative logarithm of the likelihood, $\mathcal{L}(y_i, \beta_0 + \boldsymbol{\beta}^{\mathbf{T}} \mathbf{x_i})$. Maximize likelihood is to minimize $l(y_i, \beta_0 + \boldsymbol{\beta}^{\mathbf{T}} \mathbf{x_i})$.

Parameter α decides the penalty, i.e, between L_2 ($\alpha = 0$) and L_1 ($\alpha = 1$). λ controls the weight of the whole penalty item. The higher λ is, the more weight the penalty carries comparing to likelihood. As discussed above, the ridge penalty shrinks the coefficients towards 0 but can't be exactly 0. The lasso penalty can set 0 estimates so it has the property of feature selection. The elastic net combines both. Here we have two tuning parameters, α and λ.

10.4.1 Introduction to glmnet Package

glmnet is a package that fits a penalized generalized linear model using *cyclical coordinate descent*. It successively optimizes the objective function over each parameter with others fixed, and cycles repeatedly until convergence. Since the linear model is a special case of the generalized linear model, `glmnet` can also fit a penalized linear model. Other than that, it can also fit penalized logistic regression, multinomial, Poisson, and Cox regression models.

The default family option in the function `glmnet()` is `gaussian`. It is the linear regression we discussed so far in this chapter. But the parameterization is a little different in the generalized linear model framework (we have α and λ). Let's start from our previous example, using the same training data but `glmnet()` to fit model:

```
dat <- read.csv("http://bit.ly/2P5gTw4")
# data cleaning: delete wrong observations with expense < 0
dat <- subset(dat, store_exp > 0 & online_exp > 0)
# get predictors
trainx <- dat[, grep("Q", names(dat))]
# get response
trainy <- dat$store_exp + dat$online_exp
glmfit = glmnet::glmnet(as.matrix(trainx), trainy)
```

The object `glmfit` returned by `glmnet()` has the information of the fitted model for the later operations. An easy way to extract the components is through various functions on `glmfit`, such as `plot()`, `print()`, `coef()` and `predict()`. For example, the following code visualizes the path of coefficients as penalty increases:

```
plot(glmfit, label = T)
```

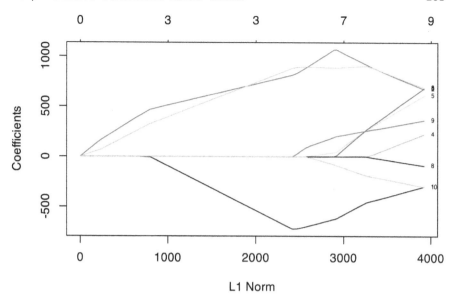

Each curve in the plot represents one predictor. The default setting is $\alpha = 1$ which means there is only lasso penalty. From left to right, L_I norm is increasing which means λ is decreasing. The bottom x-axis is L_1 norm (i.e. $\| \beta \|_1$). The upper x-axis is the effective degrees of freedom (df) for the lasso. You can check the detail for every step by:

```
print(glmfit)
```

```
Call:  glmnet(x = as.matrix(trainx), y = trainy)
```

	Df	%Dev	Lambda
1	0	0.000	3040
2	2	0.104	2770
3	2	0.192	2530
4	2	0.265	2300
5	3	0.326	2100
6	3	0.389	1910
7	3	0.442	1740
8	3	0.485	1590

```
9   3 0.521   1450
...
```

The first column Df is the degree of freedom (i.e. the number of non-zero coefficients), %Dev is the percentage of deviance explained and Lambda is the value of tuning parameter λ. By default, the function will try 100 different values of λ. However, if as λ changes, the %Dev doesn't change sufficiently, the algorithm will stop before it goes through all the values of λ. We didn't show the full output above. But it only uses 68 different values of λ. You can also set the value of λ using s= :

```
coef(glmfit, s = 1200)
```

```
## 11 x 1 sparse Matrix of class "dgCMatrix"
##                    s1
## (Intercept) 2255.2
## Q1             -390.9
## Q2              653.6
## Q3              624.4
## Q4                  .
## Q5                  .
## Q6                  .
## Q7                  .
## Q8                  .
## Q9                  .
## Q10                 .
```

When $\lambda = 1200$, there are three coefficients with non-zero estimates(Q1, Q2 and Q3). You can apply models with different values of tuning parameter to new data using predict():

```
newdat = matrix(sample(1:9, 30, replace = T), nrow = 3)
predict(glmfit, newdat, s = c(1741, 2000))
```

```
##            s1   s2
## [1,] 6004 5968
## [2,] 7101 6674
## [3,] 9158 8411
```

Each column corresponds to a value of λ. To tune the value of λ, we can easily use `cv.glmnet()` function to do cross-validation. `cv.glmnet()` returns the cross-validation results as a list object. We store the object in `cvfit` and use it for further operations.

```
cvfit = cv.glmnet(as.matrix(trainx), trainy)
```

We can plot the object using `plot()`. The red dotted line is the cross-validation curve. Each red point is the cross-validation mean squared error for a value of λ. The grey bars around the red points indicate the upper and lower standard deviation. The two gray dotted vertical lines represent the two selected values of λ, one gives the minimum mean cross-validated error (`lambda.min`), the other gives the error that is within one standard error of the minimum (`lambda.1se`).

```
plot(cvfit)
```

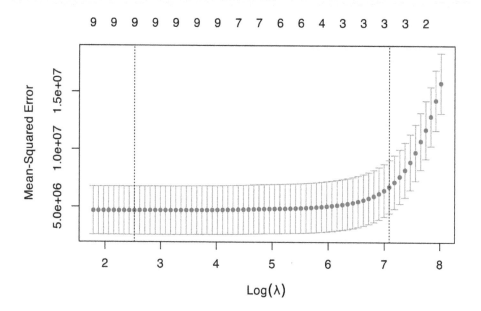

You can check the two selected λ values by:

```
# lambda with minimum mean cross-validated error
cvfit$lambda.min
```

```
## [1] 12.57
```

```
# lambda with one standard error of the minimum
cvfit$lambda.1se
```

```
## [1] 1200
```

You can look at the coefficient estimates for different λ by:

```
# coefficient estimates for model with the error
# that is within one standard error of the minimum
coef(cvfit, s = "lambda.1se")
```

```
## 11 x 1 sparse Matrix of class "dgCMatrix"
##                     s1
## (Intercept) 2255.3
## Q1              -391.1
## Q2               653.7
## Q3               624.5
## Q4                  .
## Q5                  .
## Q6                  .
## Q7                  .
## Q8                  .
## Q9                  .
## Q10                 .
```

10.4.2 Penalized Logistic Regression

10.4.2.1 Multivariate Logistic Regression Model

Logistic regression is a traditional statistical method for a two-category classification problem. It is simple yet useful. Here we use the swine disease breakout data as an example to illustrate the learning method and code implementation. Refer to section 3.2 for more details about the dataset. The goal is to predict if a farm will have a swine disease outbreak (i.e build a risk scoring system).

Consider risk scoring system construction using a sample of n observations, with information collected for G categorical predictors and one binary response variable for each observation. The predictors are 120 survey questions (i.e. $G = 120$). There were three possible answers for each question (A, B and C). So each predictor is encoded to two dummy variables (we consider C as the baseline.). Let $\mathbf{x_{i,g}}$ be the vector of dummy variables associated with the gth categorical predictor for the ith observation, where $i = 1, ..., n$, $g = 1, ..., G$. For example, if the first farm chooses B for question 2, then the corresponding observation is $\mathbf{x_{12}} = (0,1)^T$. Each question has a degree of freedom of 2.

We denote the degrees of freedom of the gth predictor by df_g, which is also the length of vector $\mathbf{x_{i,g}}$. Let y_i ($= 1$, diseased; or 0, not diseased) be the binary response for the ith observation. Denote the probability of disease for ith subject by θ_i, the model can be formulated as:

$$y_i \sim Bounoulli(\theta_i)$$

$$log\left(\frac{\theta_i}{1-\theta_i}\right) = \eta_\beta(x_i) = \beta_0 + \sum_{g=1}^{G} \mathbf{x_{i,g}}^T \boldsymbol{\beta_g}$$

where β_0 is the intercept and $\boldsymbol{\beta_g}$ is the parameter vector corresponding to the gth predictor. As we mentioned, here $\boldsymbol{\beta_g}$ has length 2.

Traditional estimation of logistic parameters $\beta = (\beta_0^T, \beta_1^T, \beta_2^T, ..., \beta_G^T)^T$ is done through maximizing the log-likelihood

$$l(\beta) \ = \ log[\prod_{i=1}^{n} \theta_i^{y_i}(1-\theta_i)^{1-y_i}]$$

$$= \ \sum_{i=1}^{n}\{y_i log(\theta_i) + (1-y_i)log(1-\theta_i)\}$$

$$= \ \sum_{i=1}^{n}\{\ y_i \eta_\beta(\mathbf{x_i}) - log[1+exp(\eta_\beta(\mathbf{x_i}))]\ \}$$

For logistic regression analysis with a large number of explanatory variables, complete- or quasi-complete-separation may lead to unstable maximum likelihood estimates as described in (Wedderburn, 1976) and (Albert and Anderson, 1984). For example:

```
library(MASS)
dat <- read.csv("http://bit.ly/2KXb1Qi")
fit <- glm(y~., dat, family = "binomial")
```

```
## Warning: glm.fit: algorithm did not converge
```

```
## Warning: glm.fit: fitted probabilities numerically 0 or
## 1 occurred
```

There is an error saying "`algorithm did not converge`." It is because there is complete separation. It happens when there are a large number of explanatory variables which makes the estimation of the coefficients unstable. To stabilize the estimation of parameter coefficients, one popular approach is the lasso algorithm with L_1 norm penalty proposed by (Tibshirani, 1996). Because the lasso algorithm can estimate some variable coefficients to be 0, it can also be used as a variable selection tool.

10.4.2.2 Penalized Logistic Regression

Penalized logistic regression adds penalty to the likelihood function:

$$\sum_{i=1}^{n}\{\ y_i \eta_\beta(\mathbf{x_i}) - log[1+exp(\eta_\beta(\mathbf{x_i}))]\ \} + \lambda(1-\alpha)\frac{\|\ \beta\ \|_2^2}{2} + \alpha\ \|\ \beta\ \|_1]$$

```
dat <- read.csv("http://bit.ly/2KXb1Qi")
trainx = dplyr::select(dat, -y)
trainy = dat$y
fit <- glmnet(as.matrix(trainx), trainy, family = "binomial")
```

The error message is gone when we use penalized regression. We can visualize the shrinking path of coefficients as penalty increases. The use of `predict()` function is a little different. For the generalized linear model, you can return different results by setting the `type` argument. The choices are

- `link`: return the link function value
- `response`: return the probability
- `class`: return the category $(0/1)$
- `coefficients`: return the coefficient estimates
- `nonzero`: return an indicator for non-zero estimates (i.e. which variables are selected)

The default setting is to predict the probability of the second level of the response variable. For example, the second level of the response variable for `trainy` here is "1":

```
levels(as.factor(trainy))
```

```
## [1] "0" "1"
```

So the model is to predict the probability of outcome "1". Take a baby example of 3 observations and 2 values of λ to show the usage of `predict()` function:

```
newdat = as.matrix(trainx[1:3, ])
predict(fit, newdat, type = "link", s = c(2.833e-02, 3.110e-02))
```

```
##          s1      s2
## 1   0.1943  0.1443
## 2  -0.9913 -1.0077
## 3  -0.5841 -0.5496
```

The first column of the above output is the predicted link function value when $\lambda = 0.02833$. The second column of the output is the predicted link function when $\lambda = 0.0311$.

Similarly, you can change the setting for `type` to produce different outputs. You can use the `cv.glmnet()` function to tune parameters. The parameter setting is nearly the same as before, the only difference is the setting of `type.measure`. Since the response is categorical, not continuous, we have different performance measurements. The most common settings of `type.measure` for classification are

- `class`: error rate
- `auc`: it is the area under the ROC for the dichotomous problem

For example:

```
cvfit = cv.glmnet(as.matrix(trainx), trainy,
                  family = "binomial", type.measure = "class")
plot(cvfit)
```

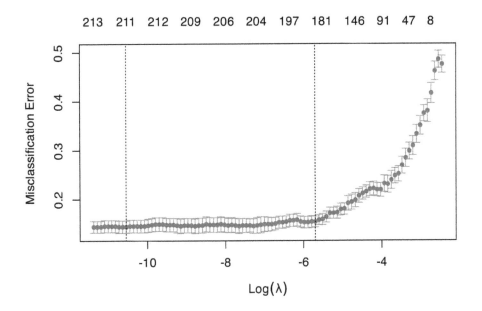

The code above uses error rate as performance criteria and use 10-fold cross-validation. Similarly, you can get the λ value for the minimum error rate and the error rate that is 1 standard error from the minimum:

```
cvfit$lambda.min
```

```
## [1] 2.643e-05
```

```
cvfit$lambda.1se
```

```
## [1] 0.003334
```

You can use the same way to get the parameter estimates and make prediction.

10.4.2.3 Group Lasso Logistic Regression

For models with categorical survey questions (explanatory variables), however, the original lasso algorithm only selects individual dummy variables instead of sets of the dummy variables grouped by the question in the survey. Another disadvantage of applying lasso to grouped variables is that the estimates are affected by the way dummy variables are encoded. Thus the group lasso (Yuan and Lin, 2007) method has been proposed to enable variable selection in linear regression models on groups of variables, instead of on single variables. For logistic regression models, the group lasso algorithm was first studied by (Kim et al., 2006). They proposed a gradient descent algorithm to solve the corresponding constrained problem, which does, however, depend on unknown constants. Meier et al. (2008) proposed a new algorithm that could work directly on the penalized problem and its convergence property does not depend on unknown constants. The algorithm is especially suitable for high-dimensional problems. It can also be applied to solve the corresponding convex optimization problem in generalized linear models. The group lasso estimator proposed by (Meier et al., 2008) for logistic regression has been shown to be statistically consistent, even with a large number of categorical predictors.

In this section, we illustrate how to use the logistic group lasso algorithm to construct risk scoring systems for predicting disease. Instead of maximizing the log-likelihood in the maximum likelihood method, the logistic group lasso estimates are calculated by minimizing the convex function:

$$S_\lambda(\boldsymbol{\beta}) = -l(\boldsymbol{\beta}) + \lambda \sum_{g=1}^{G} s(df_g) \parallel \boldsymbol{\beta_g} \parallel_2$$

where λ is a tuning parameter for the penalty and $s(\cdot)$ is a function to rescale the penalty. In lasso algorithms, the selection of λ is usually determined by cross-validation using data. For $s(\cdot)$, we use the square root function $s(df_g) = df_g^{0.5}$ as suggested in (Meier et al., 2008). It ensures the penalty is of the order of the number of parameters df_g as used in (Yuan and Lin, 2007).

Here we consider selection of the tuning parameter λ from a multiplicative grid of 100 values $\{0.96\lambda_{max}, 0.96^2\lambda_{max}, 0.96^3\lambda_{max}, ..., 0.96^{100}\lambda_{max}\}$. Here λ_{max} is defined as

$$\lambda_{max} = \max_{g \in \{1,...,G\}} \left\{ \frac{1}{s(df_g)} \parallel \mathbf{x_g}^T(\mathbf{y} - \bar{\mathbf{y}}) \parallel_2 \right\}, \qquad (10.4)$$

such that when $\lambda = \lambda_{max}$, only the intercept is in the model. When λ goes to 0, the model is equivalent to ordinary logistic regression.

Three criteria may be used to select the optimal value of λ. One is AUC which you should have seem many times in this book by now. The log-likelihood score used in (Meier et al., 2008) is taken as the average of log-likelihood of the validation data over all cross-validation sets. Another one is the maximum correlation coefficient in Yeo and Burge (Yeo and Burge, 2004) that is defined as:

$$\rho_{max} = max\{\rho_\tau | \tau \in (0, 1)\},$$

where $\tau \in (0,1)$ is a threshold to classify the predicted probability into binary disease status and ρ_τ is the Pearson correlation coefficient between the true binary disease status and the predictive disease status with threshold τ.

You can use the following package to implement the model. Install the package using:

```
devtools::install_github("netlify/NetlifyDS")
```

Load the package:

```
library("NetlifyDS")
```

The package includes the swine disease breakout data and you can load the data by:

```
data("sim1_da1")
```

You can use `cv_glasso()` function to tune the parameters:

```
# the last column of sim1_da1 response variable y
# trainx is the explanatory variable matrix
trainx = dplyr::select(sim1_da1, -y)
# save response variable as as trainy
trainy = sim1_da1$y
# get the group indicator
index <- gsub("\\..*", "", names(trainx))
```

Dummy variables from the same question are in the same group:

```
index[1:50]
```

```
##  [1] "Q1"  "Q1"  "Q2"  "Q2"  "Q3"  "Q3"  "Q4"  "Q4"
##  [9] "Q5"  "Q5"  "Q6"  "Q6"  "Q7"  "Q7"  "Q8"  "Q8"
## [17] "Q9"  "Q9"  "Q10" "Q10" "Q11" "Q11" "Q12" "Q12"
## [25] "Q13" "Q13" "Q14" "Q14" "Q15" "Q15" "Q16" "Q16"
## [33] "Q17" "Q17" "Q18" "Q18" "Q19" "Q19" "Q20" "Q20"
## [41] "Q21" "Q21" "Q22" "Q22" "Q23" "Q23" "Q24" "Q24"
## [49] "Q25" "Q25"
```

Set a series of tuning parameter values. nlam is the number of values we want to tune. It is the parameter m in $\{0.96\lambda_{max}, 0.96^2\lambda_{max}, 0.96^3\lambda_{max}, ..., 0.96^m\lambda_{max}\}$. The tuning process returns a long output and we will not report all:

```
# Tune over 100 values
nlam <- 100
# set the type of prediction
# - `link`: return the predicted link function
# - `response`: return the predicted probability
# number of cross-validation folds
kfold <- 10
cv_fit <- cv_glasso(trainx, trainy,
                    nlam = nlam, kfold = kfold, type = "link")
# only show part of the results
str(cv_fit)
```

Here we only show part of the output:

```
...
$ auc               : num [1:100] 0.573 0.567 0.535 ...
$ log_likelihood    : num [1:100] -554 -554 -553 ...
$ maxrho            : num [1:100] -0.0519 0.00666  ...
$ lambda.max.auc    : Named num [1:2] 0.922 0.94
  ..- attr(*, "names")= chr [1:2] "lambda" "auc"
$ lambda.1se.auc    : Named num [1:2] 16.74 0.81
  ..- attr(*, "names")= chr [1:2] "" "se.auc"
```

```
$ lambda.max.loglike: Named num [1:2] 1.77 -248.86
 ..- attr(*, "names")= chr [1:2] "lambda" "loglike"
$ lambda.1se.loglike: Named num [1:2] 9.45 -360.13
 ..- attr(*, "names")= chr [1:2] "lambda" "se.loglike"
$ lambda.max.maxco  : Named num [1:2] 0.922 0.708
 ..- attr(*, "names")= chr [1:2] "lambda" "maxco"
$ lambda.1se.maxco  : Named num [1:2] 14.216 0.504
 ..- attr(*, "names")= chr [1:2] "lambda" "se.maxco"
```

In the returned results:

- $ auc: the AUC values
- $ log_likelihood: log-likelihood
- $ maxrho: maximum correlation coefficient
- $ lambda.max.auc: the max AUC and the corresponding value of λ
- $ lambda.1se.auc: one standard error to the max AUC and the corresponding λ
- $ lambda.max.loglike: max log-likelihood and the corresponding λ
- $ lambda.1se.loglike: one standard error to the max log-likelihood and the corresponding λ
- $ lambda.max.maxco: maximum correlation coefficient and the corresponding λ
- $ lambda.1se.maxco: one standard error to the maximum correlation coefficient and the corresponding λ

The most common criterion is AUC. You can compare the selections from different criteria. If they all point to the same value of the tuning parameter, you can have more confidence about the choice. If they suggest very different values, then you need to concern if the tuning process is stable. You can visualize the cross validation result:

```
plot(cv_fit)
```

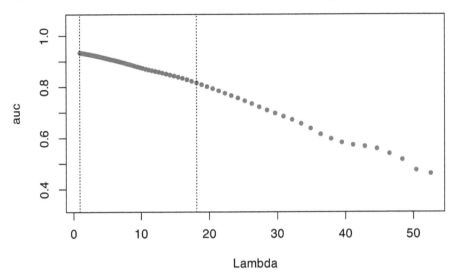

The x-axis is the value of the tuning parameter, the y-axis is AUC. The two dash lines are the value of λ for max AUC and the value for the one standard deviation to the max AUC. Once you choose the value of the tuning parameter, you can use `fitglasso()` to fit the model. For example, we can fit the model using the parameter value that gives the max AUC, which is $\lambda = 0.922$:

```
fitgl <- fitglasso(trainx, trainy,
                   lambda = 0.922, na_action = na.pass)
```

```
Lambda: 0.922  nr.var: 229
```

You can use `coef()` to get the estimates of coefficients:

```
coef(fitgl)
```

```
                  0.922
Intercept -5.318e+01
Q1.A       1.757e+00
Q1.B       1.719e+00
Q2.A       2.170e+00
Q2.B       6.939e-01
Q3.A       2.102e+00
Q3.B       1.359e+00
...
```

Use `predict_glasso()` to predict new samples:

```
prey <- predict_glasso(fitgl, trainx)
```

11

Tree-Based Methods

Tree-based models such as random forest and gradient boosted trees are frequent winners in data challenges and competitions which use standard numerical and categorical datasets. These methods, in general, provide a good baseline for model performance. This chapter describes the fundamentals of tree-based models and provides a set of standard modeling procedures.

Load R packages:

```
# install packages from CRAN
p_needed <- c('rpart', 'caret', 'partykit',
              'pROC', 'dplyr', 'ipred',
              'e1071', 'randomForest', 'gbm')

packages <- rownames(installed.packages())
p_to_install <- p_needed[!(p_needed %in% packages)]
if (length(p_to_install) > 0) {
    install.packages(p_to_install)
}

lapply(p_needed, require, character.only = TRUE)
```

11.1 Tree Basics

The tree-based models can be used for regression and classification. The goal is to stratify or segment the predictor space into a number of sub-regions. For a given observation, use the mean (regression)

DOI: 10.1201/9781351132916-11

or the mode (classification) of the training observations in the sub-region as the prediction. Tree-based methods are conceptually simple yet powerful. This type of model is often referred to as Classification And Regression Trees (CART). They are popular tools for many reasons:

1. Do not require user to specify the form of the relationship between predictors and response
2. Do not require (or if they do, very limited) data prepro-cessing and can handle different types of predictors (sparse, skewed, continuous, categorical, etc.)
3. Robust to co-linearity
4. Can handle missing data
5. Many pre-built packages make implementation as easy as a button push

CART can refer to the tree model in general, but most of the time, it represents the algorithm initially proposed by Breiman (Breiman et al., 1984). After Breiman, there are many new algo-rithms, such as ID3, C4.5, and C5.0. C5.0 is an improved version of C4.5, but since C5.0 is not open source, the C4.5 algorithm is more popular. C4.5 was a major competitor of CART. But now, all those seem outdated. The most popular tree models are Random Forest (RF) and Gradient Boosting Machine (GBM). Despite being out of favor in application, it is important to understand the mechanism of the basic tree algorithm. Because the later models are based on the same foundation.

The original CART algorithm targets binary classification, and the later algorithms can handle multi-category classification. A single tree is easy to explain but has poor accuracy. More compli-cated tree models, such as RF and GBM, can provide much better prediction at the cost of explainability. As the model becoming more complicated, it is more like a black-box which makes it very difficult to explain the relationship among predictors. There is always a trade-off between explainability and predictability.

The reason why it is called "tree" is of course because the structure has similarities. But the direction of the decision tree is opposite of a real tree, the root is on the top, and the leaf is on the

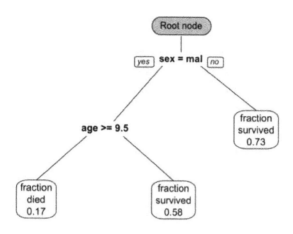

FIGURE 11.1

This is a classification tree trained from passenger survival data from the Titanic. Survival probability is predicted using sex and age.

bottom (figure 11.1). From the root node, a decision tree divides to different branches and generates more nodes. The new nodes are child nodes, and the previous node is the parent node. At each child node, the algorithm will decide whether to continue dividing. If it stops, the node is called a leaf node. If it continues, then the node becomes the new parent node and splits to produce the next layer of child nodes. At each non-leaf node, the algorithm needs to decide if it will split into branches. A leaf node contains the final "decision" on the sample's value. Here are the important definitions in the tree model:

- **Classification tree**: the outcome is discrete
- **Regression tree**: the outcome is continuous (e.g. the price of a house)
- **Non-leaf node (or split node)**: the algorithm needs to decide a split at each non-leaf node (eg: age $>= 9.5$)
- **Root node**: the beginning node where the tree starts
- **Leaf node (or Terminal node)**: the node stops splitting. It has the final decision of the model

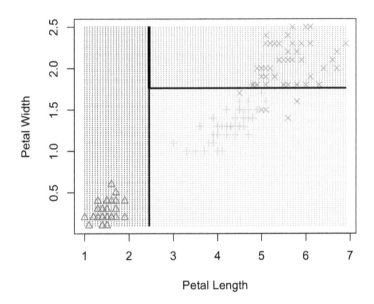

FIGURE 11.2
Example of decision regions for different types of flowers based
on petal length and width. Three different types of flowers are
classified using decisions about length and width.

- **Degree of the node**: the number of subtrees of a node
- **Degree of the tree**: the maximum degree of a node in the
  tree
- **Pruning**: remove parts of the tree that do not provide power
  to classify instances
- **Branch (or Subtree)**: the whole part under a non-leaf node
- **Child node**: the node directly after and connected to another
  node
- **Parent node**: the converse notion of a child

A single tree is easy to explain, but it can be very non-robust,
which means a slight change in the data can significantly change
the fitted tree. The predictive accuracy is not as good as other
regression and classification approaches in this book since a series
of rectangular decision regions defined by a single tree is often too
naive to represent the relationship between the dependent variable
and the predictors. Figure 11.2 shows an example of decision regions

based on the iris data in R where we use petal length (Petal.Length) and width (Petal.Width) to decide the type of flowers (Species). You can get the data frame using the following code:

```
data("iris")
head(iris)
```

```
##    Sepal.Length Sepal.Width Petal.Length Petal.Width
## 1          5.1         3.5          1.4         0.2
## 2          4.9         3.0          1.4         0.2
## 3          4.7         3.2          1.3         0.2
## 4          4.6         3.1          1.5         0.2
## 5          5.0         3.6          1.4         0.2
## 6          5.4         3.9          1.7         0.4
##    Species
## 1  setosa
## 2  setosa
## 3  setosa
## 4  setosa
## 5  setosa
## 6  setosa
```

To overcome these shortcomings, researchers have proposed ensemble methods which combine many trees. Ensemble tree models typically have much better predictive performance than a single tree. We will introduce those models in later sections.

11.2 Splitting Criteria

The splitting criteria used by the regression tree and the classification tree are different. Like the regression tree, the goal of the classification tree is to divide the data into smaller, more homogeneous groups. Homogeneity means that most of the samples at each node are from one class. The original CART algorithm uses Gini impurity as the splitting criterion; The later ID3, C4.5, and

C5.0 use entropy. We will look at three most common splitting criteria.

11.2.1 Gini Impurity

Gini impurity (Breiman et al., 1984) is a measure of non-homogeneity. It is widely used in classification tree. It is defined as:

$$Gini\ Index = \Sigma_i p_i(1 - p_i)$$

where p_i is the probability of class i and the interval of Gini is $[0, 0.5]$. For a two-class problem, the Gini impurity for a given node is:

$$p_1(1 - p_1) + p_2(1 - p_2)$$

It is easy to see that when the sample set is pure, one of the probability is 0 and the Gini score is the smallest. Conversely, when $p_1 = p_2 = 0.5$, the Gini score is the largest, in which case the purity of the node is the smallest. Let's look at an example. Suppose we want to determine which students are computer science (CS) majors. Here is the simple hypothetical classification tree result obtained with the gender variable.

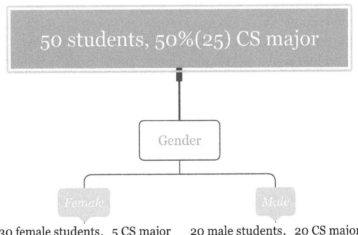

30 female students, 5 CS major   20 male students, 20 CS major

Let's calculate the Gini impurity for splitting node "Gender":

1. Gini impurity for "Female" $= \frac{1}{6} \times \frac{5}{6} + \frac{5}{6} \times \frac{1}{6} = \frac{5}{18}$
2. Gini impurity for "Male" $= 0 \times 1 + 1 \times 0 = 0$

The Gini impurity for the node "Gender" is the following weighted average of the above two scores:

$$\frac{3}{5} \times \frac{5}{18} + \frac{2}{5} \times 0 = \frac{1}{6}$$

The Gini impurity for the 50 samples in the parent node is $\frac{1}{2}$. It is easy to calculate the Gini impurity drop from $\frac{1}{2}$ to $\frac{1}{6}$ after splitting. The split using "gender" causes a Gini impurity decrease of $\frac{1}{3}$. The algorithm will use different variables to split the data and choose the one that causes the most substantial Gini impurity decrease.

11.2.2 Information Gain (IG)

Looking at the samples in the following three nodes, which one is the easiest to describe? It is obviously C. Because all the samples in C are of the same type, so the description requires the least amount of information. On the contrary, B needs more information, and A needs the most information. In other words, C has the highest purity, B is the second, and A has the lowest purity. We need less information to describe nodes with higher purity.

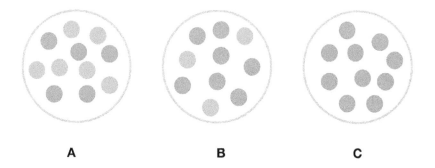

A   B   C

A measure of the degree of disorder is entropy which is defined as:

$$Entropy = -\Sigma_i p_i log_2(p_i)$$

where p_i is the probability of class i and the interval of entropy is $[0, 1]$. For a two-class problem:

$$Entropy = -plog_2 p - (1 - p)log_2(1 - p)$$

where p is the percentage of one type of samples. If all the samples in one node are of one type (such as C), the entropy is 0. If the proportion of each type in a node is 50%, the entropy is 1. We can use entropy as splitting criteria. The goal is to decrease entropy as the tree grows. As an analogy, entropy in physics quantifies the level of disorder and the goal here is to have the least disorder.

Similarly, the entropy of a splitting node is the weighted average of the entropy of each child. In the above tree for the students, the entropy of the root node with all 50 students is $-\frac{25}{50}log_2\frac{25}{50} - \frac{25}{50}log_2\frac{25}{50} = 1$. Here an entropy of 1 indicates that the purity of the node is the lowest, that is, each type takes up half of the samples.

The entropy of the split using variable "gender" can be calculated in three steps:

1. Entropy for "Female" $= -\frac{5}{30}log_2\frac{5}{30} - \frac{25}{30}log_2\frac{25}{30} = 0.65$
2. Entropy for "Male" $= 0 \times 1 + 1 \times 0 = 0$
3. Entropy for the node "Gender" is the weighted average of the above two entropy numbers: $\frac{3}{5} \times 0.65 + \frac{2}{5} \times 0 = 0.39$

So entropy decreases from 1 to 0.39 after the split and the IG for "Gender" is 0.61.

11.2.3 Information Gain Ratio (IGR)

ID3 uses information gain as the splitting criterion to train the classification tree. A drawback of information gain is that it is biased towards choosing attributes with many values, resulting in overfitting (selecting a feature that is non-optimal for prediction) (HSSINA et al., 2014).

To understand why let's look at another hypothetical scenario. Assume that the training set has students' birth month as a feature. You might say that the birth month should not be considered in this case because it intuitively doesn't help tell the student's major.

Yes, you're right. However, practically, we may have a much more complicated dataset, and we may not have such intuition for all the features. So, we may not always be able to determine whether a feature makes sense or not. If we use the birth month to split the data, the corresponding entropy of the node "Birth Month" is 0.24 (the sum of column "Weighted Entropy" in the table), and the information gain is 0.76, which is larger than the IG of "Gender" (0.61). So between the two features, IG would choose "Birth Month" to split the data.

Birth Month	CS Major		% CS Major	Entropy	Weight	Weighted Entropy
	Yes	No				
Jan	0	3	0%	0.00	0.06	0.00
Feb	1	0	100%	0.00	0.02	0.00
Mar	0	2	0%	0.00	0.04	0.00
Apr	2	1	67%	0.28	0.06	0.02
May	5	3	63%	0.29	0.16	0.05
Jun	4	2	67%	0.28	0.12	0.03
Jul	2	0	100%	0.00	0.04	0.00
Aug	2	3	40%	0.29	0.1	0.03
Sep	2	2	50%	0.30	0.08	0.02
Oct	2	3	40%	0.29	0.1	0.03
Nov	2	2	50%	0.30	0.08	0.02
Dec	3	4	43%	0.30	0.14	0.04

To overcome this problem, C4.5 uses the "information gain ratio" instead of "information gain." The gain ratio is defined as:

$$Gain\ Ratio = \frac{Information\ Gain}{Split\ Information}$$

where split information is:

$$Split\ Information = -\Sigma_{c=1}^{C} p_c log(p_c)$$

p_c is the proportion of samples in category c. For example, there are three students with the birth month in Jan, 6% of the total 50 students. So the p_c for "Birth Month = Jan" is 0.06. The split information measures the intrinsic information that is independent of the sample distribution inside different categories. The gain ratio corrects the IG by taking the intrinsic information of a split into account.

Feature	Entropy	Information Gain	Split Information	Gain Ratio
Birth Month	0.24	0.76	3.4	0.22
Gender	0.39	0.61	0.97	0.63

The split information for the birth month is 3.4, and the gain ratio is 0.22, which is smaller than that of gender (0.63). The gain ratio refers to use gender as the splitting feature rather than the birth month. Gain ratio favors attributes with fewer categories and leads to better generalization (less overfitting).

11.2.4 Sum of Squared Error (SSE)

The previous two metrics are for classification tree. The SSE is the most widely used splitting metric for regression. Suppose you want to divide the data set S into two groups of S_1 and S_2, where the selection of S_1 and S_2 needs to minimize the sum of squared errors:

$$SSE = \Sigma_{i \in S_1}(y_i - \bar{y}_1)^2 + \Sigma_{i \in S_2}(y_i - \bar{y}_2)^2 \qquad (11.1)$$

In equation (11.1), $\bar{y}_1$ and $\bar{y}_2$ are the average of the sample in S_1 and S_2. The way regression tree grows is to automatically decide on the splitting variables and split points that can maximize **SSE reduction**. Since this process is essentially a recursive segmentation, this approach is also called recursive partitioning.

Take a look at this simple regression tree for the height of 10 students:

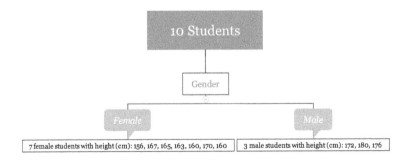

You can calculate the SSE using the following code:

1. SSE for "Female" is 136
2. SSE for "Male" is 32
3. SSE for splitting node "Gender" is the sum of the above two numbers which is 168

SSE for the 10 students in root node is 522.9. After the split, SSE decreases from 522.9 to 168.

If there is another possible way of splitting, divide it by major, as follows:

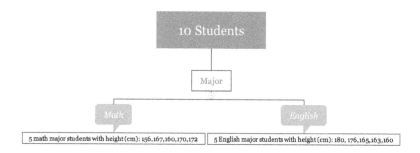

In this situation:

1. SSE for "Math" is 184
2. SSE for "English" is 302.8
3. SSE for splitting node "Major" is the sum of the above two numbers which is 486.8

Splitting data using variable "gender" reduced SSE from 522.9 to 168; using variable "major" reduced SSE from 522.9 to 486.8. Based on SSE reduction, you should use gender to split the data.

The three splitting criteria mentioned above are the basis for building a tree model.

11.3 Tree Pruning

Pruning is the process that reduces the size of decision trees. It reduces the risk of overfitting by limiting the size of the tree or removing sections of the tree that provide little power.

Limit the size

You can limit the tree size by setting some parameters.

- Minimum sample size at each node: Defining the minimum sample size at the node helps to prevent the leaf nodes having only one sample. The sample size can be a tuning parameter. If it is too large, the model tends to under-fit. If it is too small, the model tends to over-fit. In the case of severe class imbalance, the minimum sample size may need to be smaller because the number of samples in a particular class is small.

- Maximum depth of the tree: If the tree grows too deep, the model tends to over-fit. It can be a tuning parameter.

- Maximum number of terminal nodes: Limit on the terminal nodes works the same as the limit on the depth of the tree. They are proportional.

- The number of variables considered for each split: the algorithm randomly selects variables used in finding the optimal split point at each level. In general, the square root of the number of all variables works best, which is also the default setting for many functions. However, people often treat it as a tuning parameter.

Remove branches

Another way is to first let the tree grow as much as possible and then go back to remove insignificant branches. The process reduces the depth of the tree. The idea is to overfit the training set and then correct using cross-validation. There are different implementations.

- cost/complexity penalty

The idea is that the pruning minimizes the penalized error SSE_λ with a certain value of tuning parameter λ.

$$SSE_\lambda = SSE + \lambda \times (complexity)$$

Here complexity is a function of the number of leaves. For every given λ, we want to find the tree that minimizes this penalized error. Breiman presents the algorithm to solve the optimization (Breiman et al., 1984).

To find the optimal pruning tree, you need to iterate through a series of values of λ and calculate the corresponding SSE. For the same λ, SSE changes over different samples. Breiman et al. suggested using cross-validation (Breiman et al., 1984) to study the variation of SSE under each λ value. They also proposed a standard deviation criterion to give the simplest tree: within one standard deviation, find the simplest tree that minimizes the absolute error. Another method is to choose the tree size that minimizes the numerical error (Hastie et al., 2008).

- Error-based pruning

This method was first proposed by Quinlan (Quinlan, 1987). The idea behind is intuitive. All split nodes of the tree are included in the initial candidate pool. Pruning a split node means removing the entire subtree under the node and setting the node as a terminal node. The data is divided into 3 subsets for:

(1)   training a complete tree

(2)   pruning

(3)   testing the final model

You train a complete tree using the subset (1) and apply the tree on the subset (2) to calculate the accuracy. Then prune the tree based on a node and apply that on the subset (2) to calculate another accuracy. If the accuracy after pruning is higher or equal to that from the complete tree, then we set the node as a terminal node. Otherwise, keep the subtree under the node. The advantage of this method is that it is easy to compute. However, when the size of the subset (2) is much smaller than that of the subset (1), there is a risk of over-pruning. Some researchers found that this method results in more accurate trees than pruning process based on tree size (F. Espoito and Semeraro, 1997).

- Error-complexity pruning

This method is to search for a trade-off between error and complexity. Assume we have a splitting node t, and the corresponding subtree T. The error cost of the node is defined as:

$$R(t) = r(t) \times p(t) = \frac{misclassified \; sample \; size \; of \; the \; node}{total \; sample \; size}$$

where $r(t)$ is the error rate associate with the node as if it is a terminal node:

$$r(t) = \frac{misclassified \; sample \; size \; of \; the \; node}{sample \; size \; of \; the \; node}$$

$p(t)$ is the ratio of the sample of the node to the total sample:

$$p(t) = \frac{sample \; size \; of \; the \; node}{total \; sample \; size}$$

The multiplication $r(t) \times p(t)$ cancels out the sample size of the node. If we keep node t, the error cost of the subtree T is:

$$R(T) = \Sigma_{i=no. \; of \; leaves \; in \; subtree \; T} R(i)$$

The error-complexity measure of the node is:

$$a(t) = \frac{R(t) - R(T)_t}{no. \; of \; leaves - 1}$$

Based on the metrics above, the pruning process is (Patel and Upadhyay, 2012):

1. Calculate a for every node t.
2. Prune the node with the lowest value.
3. Repeat 1 and 2. It produces a pruned tree each time and they form a forest.
4. Select the tree with the best overall accuracy.

- Minimum error pruning

Niblett and Brotko introduced this pruning method in 1991 (Cestnik and Bratko, 1991). The process is a bottom-up approach which seeks a single tree that minimizes the expected error rate on new samples. If we prune a splitting point t, all the samples under t will be classified as from one category, say category c. If we prune the subtree, the expected error rate is:

$$E(t) = \frac{n_t - n_{t,c} + k - 1}{n_t + k}$$

where:

$$k = number\ of\ categories$$
$$n_t = sample\ size\ under\ node\ t$$
$$n_{t,c} = number\ of\ sample\ under\ t\ that\ belong\ to\ category\ c$$

Based on the above definition, the pruning process is (F. Espoito and Semeraro, 1997):

- Calculate the expected error rate for each non-leaf node if that subtree is pruned
- Calculate the expected error rate for each non-leaf node if that subtree is not pruned
- If pruning the node leads to higher expected rate, then keep the subtree; otherwise, prune it.

11.4   Regression and Decision Tree Basic

11.4.1   Regression Tree

Let's look at the process of building a regression tree (Gareth James and Tibshirani, 2015). There are two steps:

1. Divide predictors space – that is a set of possible values of $X_1, X_2, ..., X_p$ – into J distinct and non-overlapping regions: $R_1, R_2, ..., R_J$
2. For every observation that falls into the region R_j, the prediction is the mean of the response values for the training observations in R_j

Let's go back to the previous simple example. If we use the variable "Gender" to divide the observations, we obtain two regions R_1 (female) and R_2 (male).

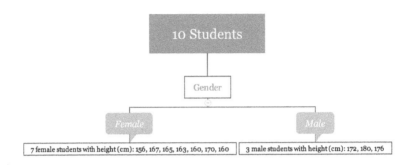

```
y1 <- c(156, 167, 165, 163, 160, 170, 160)
y2 <- c(172, 180, 176)
```

The sample average for region R_1 is 163, for region R_2 is 176. For a new observation, if it is female, the model predicts the height to be 163, if it is male, the predicted height is 176. Calculating the mean is easy. Let's look at the first step in more detail which is to divide the space into $R_1, R_2, \ldots, R_J$.

In theory, the region can be any shape. However, to simplify the problem, we divide the predictor space into high-dimensional rectangles. The goal is to divide the space in a way that minimize RSS. Practically, it is nearly impossible to consider all possible partitions of the feature space. So we use an approach named recursive binary splitting, a top-down, greedy algorithm. The process starts from the top of the tree (root node) and then successively splits the predictor space. Each split produces two branches (hence binary). At each step of the process, it chooses the best split at that particular step, rather than looking ahead and picking a split that leads to a better tree in general (hence greedy).

$$R_1(j, s) = \{X | X_j < s\} \text{ and } R_2(j, s) = \{X | X_j \geq s\}$$

Calculate the RSS decrease after the split. For different (j, s), search for the combination that minimizes the RSS, that is to minimize the following:

$$\Sigma_{i:x_i \in R_1(j,s)}(y_i - \hat{y}_{R_1})^2 + \Sigma_{i:x_i \in R_2(j,s)}(y_i - \hat{y}_{R_2})^2$$

where $\hat{y}_{R_1}$ is the mean of all samples in R_1, $\hat{y}_{R_2}$ is the mean of samples in R_2. It can be quick to optimize the equation above. Especially when p is not too large.

Next, we continue to search for the split that optimize the RSS. Note that the optimization is limited in the sub-region. The process keeps going until a stopping criterion is reaches. For example, continue until no region contains more than 5 samples or the RSS decreases less than 1%. The process is like a tree growing.

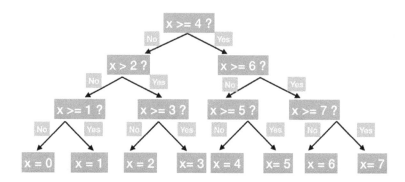

There are multiple R packages for building regression tree, such as ctree, rpart and tree. rpart is widely used for building a single tree. The split is based on CART algorithm, using rpart() function from the package. There are some parameters that controls the model fitting, such as the minimum number of observations that must exist in a node in order for a split to be attempted, the minimum number of observations in any leaf node etc. You can set those parameter using rpart.control.

A more convenient way is to use train() function in caret package. The package can call rpart() function and train the model through cross-validation. In this case, the most common parameters are cp (complexity parameter) and maxdepth (the maximum depth of any node of the final tree). To tune the complexity parameter, set `method = "rpart"`. To tune the maximum tree depth, set `method = "rpart2"`. Now let us use the customer expenditure regression example to illustrate:

```
dat <- read.csv("http://bit.ly/2P5gTw4")
# data cleaning: delete wrong observations
dat <- subset(dat, store_exp > 0 & online_exp > 0)
# use the 10 survey questions as predictors
trainx <- dat[, grep("Q", names(dat))]
# use the sum of store and online expenditure as response variable
# total expenditure = store expenditure + online expenditure
trainy <- dat$store_exp + dat$online_exp
set.seed(100)
rpartTune <- train(trainx, trainy,
                   method = "rpart2",
                   tuneLength = 10,
       trControl = trainControl(method = "cv"))
plot(rpartTune)
```

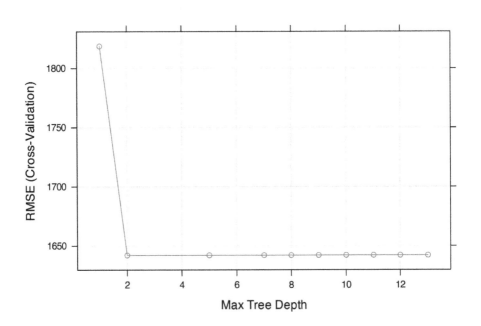

RMSE doesn't change much when the maximum is larger than 2. So we set the maximum depth to be 2 and refit the model:

```
rpartTree <- rpart(trainy ~ ., data = trainx, maxdepth = 2)
```

You can check the result using `print()`:

```
print(rpartTree)
```

```
## n= 999
##
## node), split, n, deviance, yval
##       * denotes terminal node
##
## 1) root 999 1.581e+10   3479.0
##    2) Q3< 3.5 799 2.374e+09   1819.0
##      4) Q5< 1.5 250 3.534e+06    705.2 *
##      5) Q5>=1.5 549 1.919e+09   2326.0 *
##    3) Q3>=3.5 200 2.436e+09 10110.0 *
```

You can see that the final model picks `Q3` and `Q5` to predict total expenditure. To visualize the tree, you can convert `rpart` object to `party` object using `partykit` then use `plot()` function:

```
rpartTree2 <- as.party(rpartTree)
plot(rpartTree2)
```

11.4.2 Decision Tree

Similar to a regression tree, the goal of a classification tree is to stratifying the predictor space into a number of sub-regions that are more homogeneous. The difference is that a classification tree is used to predict a categorical response rather than a continuous

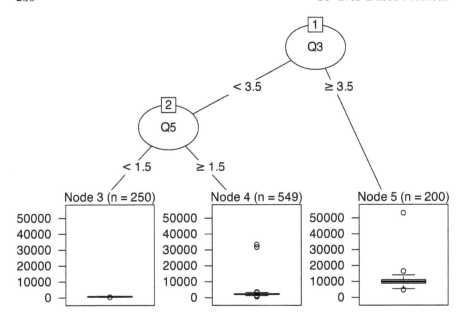

one. For a classification tree, the prediction is the most commonly occurring class of training observations in the region to which an observation belongs. The splitting criteria for a classification tree are different. The most common criteria are entropy and Gini impurity. CART uses Gini impurity and C4.5 uses entropy.

When the predictor is continuous, the splitting process is straightforward. When the predictor is categorical, the process can take different approaches:

1. Keep the variable as categorical and group some categories on either side of the split. In this way, the model can make more dynamic splits but must treat the categorical predictor as an ordered set of bits.
2. Use one-hot encoding (figure 11.3). Encode the categorical variable as a set of dummy (binary) variables. The model considers these dummy variables separately and evaluates each of these on one split point (because there are only two possible values: 0/1). This way, the information in the categorical variable is decomposed into independent bits of information.

Segment	One-hot encoding	segmentConspicuous	segmentPrice	segmentQuality	segmentStyle
Conspicuous		1	0	0	0
Price		0	1	0	0
Quality		0	0	1	0
Style		0	0	0	1

FIGURE 11.3
One-hot encoding

When fitting tree models, people need to choose the way to treat categorical predictors. If you know some of the categories have higher predictability, then the first approach may be better. In the rest of this section, we will build tree models using the above two approaches and compare them.

Let's build a classification model to identify the gender of the customer:

```
dat <- read.csv("http://bit.ly/2P5gTw4")
# use the 10 survey questions as predictors
trainx1 <- dat[, grep("Q", names(dat))]
# add a categorical predictor
# use two ways to treat categorical predictor
# trainx1: use approach 1, without encoding
trainx1$segment <- dat$segment

# trainx2: use approach 2, encode it to a set of dummy variables
dumMod <- dummyVars(
  ~.,
  data = trainx1,
  # Combine the previous variable and the level name
  # as the new dummy variable name
  levelsOnly = F
  )
trainx2 <- predict(dumMod, trainx1)
# the response variable is gender
trainy <- dat$gender
```

```
# check outcome balance
table(dat$gender) %>% prop.table()
```

```
##
## Female    Male
##  0.554   0.446
```

The outcome is pretty balanced, with 55% female and 45% male. We use `train()` function in `caret` package to call `rpart` to build the model. We can compare the model results from the two approaches:

```
CART
```

```
1000 samples
  11 predictor
   2 classes: 'Female', 'Male'
```

```
No pre-processing
Resampling: Cross-Validated (10 fold)
Summary of sample sizes: 901, 899, 900, 900, 901, 900, ...
Resampling results across tuning parameters:
```

```
  cp       ROC     Sens    Spec
  0.00000  0.6937  0.6517  0.6884
  0.00835  0.7026  0.6119  0.7355
  0.01670  0.6852  0.5324  0.8205
  0.02505  0.6803  0.5107  0.8498
  0.03340  0.6803  0.5107  0.8498
```

```
  ......
```

```
  0.23380  0.6341  0.5936  0.6745
  0.24215  0.5556  0.7873  0.3240
```

```
ROC was used to select the optimal model using the largest value.
The final value used for the model was cp = 0.00835.
```

The above keeps the variable as categorical without encoding. Here `cp` is the complexity parameter. It is used to decide when to stop growing the tree. `cp = 0.01` means the algorithm only keeps the split that improves the corresponding metric by more than 0.01. Next, let's encode the categorical variable to be a set of dummy variables and fit the model again:

```
rpartTune2 <- caret::train(
  trainx2, trainy, method = "rpart",
  tuneLength = 30,
  metric = "ROC",
  trControl = trainControl(method = "cv",
                           summaryFunction = twoClassSummary,
                           classProbs = TRUE,
                           savePredictions = TRUE)
)
```

Compare the results of the two approaches.

```
rpartRoc <- pROC::roc(response = rpartTune1$pred$obs,
              predictor = rpartTune1$pred$Female,
              levels = rev(levels(rpartTune1$pred$obs)))

rpartFactorRoc <- pROC::roc(response = rpartTune2$pred$obs,
                  predictor = rpartTune2$pred$Female,
                  levels = rev(levels(rpartTune1$pred$obs)))

plot.roc(rpartRoc,
     type = "s",
     print.thres = c(.5),
     print.thres.pch = 3,
```

```
      print.thres.pattern = "",
      print.thres.cex = 1.2,
      col = "red", legacy.axes = TRUE,
      print.thres.col = "red")

plot.roc(rpartFactorRoc,
      type = "s",
      add = TRUE,
      print.thres = c(.5),
      print.thres.pch = 16, legacy.axes = TRUE,
      print.thres.pattern = "",
      print.thres.cex = 1.2)

legend(.75, .2,
      c("Grouped Categories", "Independent Categories"),
      lwd = c(1, 1),
      col = c("black", "red"),
      pch = c(16, 3))
```

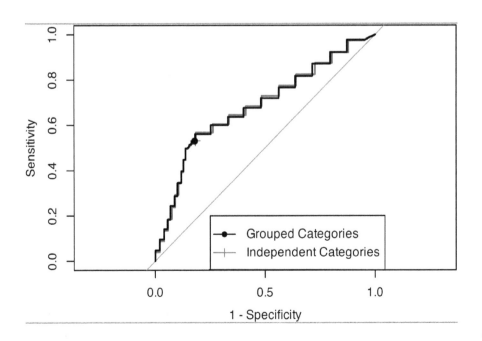

In this case, the two approaches lead to similar model performance.

Single tree is straightforward and easy to interpret but it has problems:

1. Low accuracy
2. Unstable: little change in the training data leads to very different trees.

One way to overcome those is to use an ensemble of trees. In the rest of this chapter, we will introduce three ensemble methods (combine many models' predictions): bagging tree, random forest, and gradient boosted machine. Those ensemble approaches have significant higher accuracy and stability. However, it comes with the cost of interpretability.

11.5 Bagging Tree

As mentioned before, a single tree is unstable. If you randomly separate the sample to be two parts and fit tree model on each, you can get two very different trees. A stable model should give a similar result on different random samples. Some traditional statistical models have high stability, such as linear regression. *Ensemble methods* appeared in the 1990s which can effectively stabilize the model. *Bootstrapping* is a type of process where you repeated draw samples of the same size from a single original sample with replacement (Efron and Tibshirani, 1986). *Bootstrap aggregation* (Bagged) is an ensemble technique proposed by Leo Breiman (Breiman 1996a). It uses bootstrapping in conjunction with any model to construct an ensemble. The process is very straightforward:

Assume that there are n independent random variables $Z_1, ..., Z_n$ with variance σ^2. Then the variance of the mean $\bar{Z}$ is $\frac{\sigma^2}{n}$. It is easy to see why bagged models have less variance. Since bootstrapping is to sample with replacement, it means some samples are selected multiple times and some not at all. Those left out samples are called out-of-bag. You can use the out-of-bag sample

---

Algorithm 3 Bagging tree

1: Build a model on different bootstrap samples to form an ensemble, say B samples
2: For a new sample, each model will give a prediction: $\hat{f}^1(x), \hat{f}^2(x) ..., \hat{f}^B(x)$
3: The bagged model's prediction is the average of all the predictions:
$$\hat{f}_{avg}(x) = \frac{1}{B}\Sigma_{b=1}^{B}\hat{f}^b(x)$$

---

to access the model performance. For regression, the prediction is a simple average. For classification, the prediction is the category with the most "votes." Here, the number of trees, B is a parameter you need to decide, i.e. tuning parameter. **Bagging is a general approach that can be applied to different learners. Here we only discuss in the context of decision trees.**

The advantages of bagging tree are

- Bagging stabilizes the model predictions by averaging the results. If we have 10 bootstrap samples and fit a single tree on each of those, we may get 10 trees with very different structures and leading to different predictions for a new sample. But if we use the average of the 10 predictions as the final prediction, then the result is much more stable. It means if we have another 10 samples and do it all-over again, we will get very similar averaged prediction.

- Bagging provides more accurate predictions. If the goal is to predict rather than interpret, then the ensemble approach definitely has an advantage, especially for unstable models. However, for stable models (such as regression, MARS), bagging may bring marginal improvement for the model performance.

- Bagging can use out-of-bag samples to evaluate model performance. For each model in the ensemble, we can calculate the value of the model performance metric (you can decide what metric to use). You can use the average of all the out-of-bag performance values to gauge the predictive performance of the entire ensemble. This correlates well with either cross-validation estimates or test set estimates. On average, each tree uses about 2/3 of the samples, and the rest 1/3 is used as out-of-bag. When the number of bootstrap samples is large enough, the out-of-bag performance estimate approximates that from leave one out cross-validation.

You need to choose the number of bootstrap samples. The author of "Applied Predictive Modeling" (Kuhn and Johnston, 2013) points out that often people see an exponential decrease in predictive improvement as the number of iterations increases. Most of the predictive power is from a small portion of the trees. Based on their experience, model performance can have small improvements up to 50 bagging iterations. If it is still not satisfying, they suggest trying other more powerfully predictive ensemble methods such as random forests and boosting which will be described in the following sections.

The disadvantages of bagging tree are

- As the number of bootstrap samples increases, the computation and memory requirements increase as well. You can mitigate this disadvantage by parallel computing. Since each bootstrap sample and modeling is independent of any other sample and model, you can easily parallelize the bagging process by building those models separately and bring back the results in the end to generate the prediction.

- The bagged model is difficult to explain which is common for all ensemble approaches. However, you can still get variable importance by combining measures of importance across the ensemble. For example, we can calculate the RSS decrease for each variable across all trees and use the average as the measurement of the importance.

- Since the bagging tree uses all of the original predictors as everey split of every tree, those trees are related with each other. The tree correlation prevents bagging from optimally reducing the variance of the predicted values. See (Hastie et al., 2008) for a mathematical illustration of the tree correlation phenomenon.

Let's look at how to use R to build bagging tree using survey question to predict customer gender based on the customer dataset. Get the predictors and response variable first:

```
dat <- read.csv("http://bit.ly/2P5gTw4")
# use the 10 survey questions as predictors
trainx <- dat[, grep("Q", names(dat))]
# add segment as a predictor
# don't need to encode it to dummy variables
trainx$segment <- as.factor(dat$segment)
# use gender as the response variable
trainy <- as.factor(dat$gender)
```

Then fit the model using train function in caret package. Here we just set the number of trees to be 1000. You can tune that parameter.

```
set.seed(100)
bagTune <- caret::train(trainx, trainy,
                    method = "treebag",
                    nbagg = 1000,
                    metric = "ROC",
                    trControl = trainControl(method = "cv",
                    summaryFunction = twoClassSummary,
                    classProbs = TRUE,
                    savePredictions = TRUE))
```

The model results are

```
bagTune
```

```
## Bagged CART
##
## 1000 samples
##    11 predictor
##     2 classes: 'Female', 'Male'
##
## No pre-processing
## Resampling: Cross-Validated (10 fold)
## Summary of sample sizes: 901, 899, 900, 900, 901, 900, ...
## Resampling results:
##
##    ROC      Sens     Spec
##    0.7093   0.6533   0.6774
```

Since we only have a handful of variables in this example, the maximum AUC doesn't improve by using bagging tree. But it makes a difference when we have more predictors.

11.6 Random Forest

Since the tree correlation prevents bagging from optimally reducing the variance of the predicted values, a natural way to improve the model performance is to reduce the correlation among trees. That is what random forest aims to do: improve the performance of bagging by de-correlating trees.

From a statistical perspective, you can de-correlate trees by introducing randomness when you build each tree. One approach (Ho, 1998; Amit and Geman, 1997) is to randomly choose m variables to use each time you build a tree. Dietterich (2000) came up with the idea of random split selection which is to randomly choose m variables to use at each splitting node. Based on the different generalizations to the original bagging algorithm, Breiman (Breiman, 2001a) came up with a unified algorithm called *random forest*.

When building a tree, the algorithm randomly chooses m variables to use at each splitting node. Then choose the best one out of the m to use at that node. In general, people use $m = \sqrt{p}$.

For example, if we use 10 questions from the questionnaire as predictors, then at each node, the algorithm will randomly choose 4 candidate variables. Since those trees in the forest don't always use the same variables, tree correlation is less than that in bagging. It tends to work better when there are more predictors. Since we only have 10 predictors here, the improvement from the random forest is marginal. The number of randomly selected predictors is a tuning parameter in the random forest. Since random forest is computationally intensive, we suggest starting with value around $m = \sqrt{p}$. Another tuning parameter is the number of trees in the forest. You can start with 1000 trees and then increase the number until performance levels off. The basic random forest is shown in Algorithm 4.

Algorithm 4 Random forest

1: Select the number of trees, B
2: **for** i=1 to B **do**
3:     Generate a bootstrap sample of the original data
4:     Train a tree on this sample
5:     **for** each split **do**
6:         Randomly select m ($<$ p) predictors
7:         Choose the best one out of the m and partition the data
8:     **end for**
9:     Use typical tree model stopping criteria to determine when a tree is complete without pruning
10: **end for**

When $m = p$, random forest is equal to the bagging tree. When the predictors are highly correlated, then smaller m tends to work better. Let's use the caret package to train a random forest:

```
# tune across a list of numbers of predictors
mtryValues <- c(1:5)
set.seed(100)
rfTune <- train(x = trainx,
                y = trainy,
```

```
                    # set the model to be random forest
                    method = "rf",
                    ntree = 1000,
                    tuneGrid = data.frame(.mtry = mtryValues),
                    importance = TRUE,
                    metric = "ROC",
                    trControl = trainControl(method = "cv",
                               summaryFunction = twoClassSummary,
                               classProbs = TRUE,
                               savePredictions = TRUE))
```

```
rfTune
```

```
## Random Forest
##
## 1000 samples
##   11 predictor
##    2 classes: 'Female', 'Male'
##
## No pre-processing
## Resampling: Cross-Validated (10 fold)
## Summary of sample sizes: 899, 900, 900, 899, 899, 901, ...
## Resampling results across tuning parameters:
##
##   mtry  ROC     Sens    Spec
##   1     0.7169  0.5341  0.8205
##   2     0.7137  0.6334  0.7175
##   3     0.7150  0.6478  0.6995
##   4     0.7114  0.6550  0.6950
##   5     0.7092  0.6514  0.6882
##
## ROC was used to select the optimal model using
##  the largest value.
## The final value used for the model was mtry = 1.
```

In this example, since the number of predictors is small, the result of the model indicates that the optimal number of candidate variables at each node is 1. The optimal AUC is not too much higher than that from bagging tree.

If you have selected the values of tuning parameters, you can also use the randomForest package to fit a random forest.

```
rfit = randomForest(trainy ~ ., trainx, mtry = 1, ntree = 1000)
```

Since bagging tree is a special case of random forest, you can fit the bagging tree by setting $mtry = p$. Function importance() can return the importance of each predictor:

```
importance(rfit)
```

```
##              MeanDecreaseGini
## Q1                      9.056
## Q2                      7.582
## Q3                      7.611
## Q4                     12.308
## Q5                      5.628
## Q6                      9.740
## Q7                      6.638
## Q8                      7.829
## Q9                      5.955
## Q10                     4.781
## segment                11.185
```

You can use varImpPlot() function to visualize the predictor importance:

```
varImpPlot(rfit)
```

It is easy to see from the plot that segment and Q4 are the top two variables to classify gender.

rfit

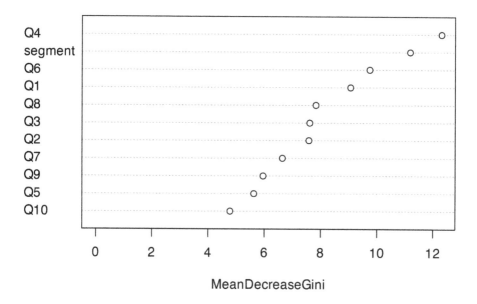

11.7 Gradient Boosted Machine

Boosting models were developed in the 1980s (Valiant, 1984; Michael and Leslie, 1989) and were originally for classification problems. Due to the excellent model performance, they were widely used for a variety of applications, such as gene expression (Dudoit et al., 2002; Ben-Dor et. al, 2000), chemical substructure classification (Varmuza et al., 2003), music classification (Bergstra et al., 2006), etc. The first effective implementation of boosting is Adaptive Boosting (AdaBoost) algorithm came up by (Yoav and Robert, 1999). After that, some researchers (Friedman et al., 2000) started to connect the boosting algorithm with some statistical concepts, such as loss function, additive model, logistic regression. Friedman pointed out that boosting can be considered as a forward stagewise additive model that minimizes exponential loss. The new view of boosting in a statistical framework enabled the method to be extended to regression problems.

The idea is to combine a group of weak learners (a classifier that is marginally better than random guess) to produce a strong learner. Like bagging, boosting is a general approach that can be applied to different learners. Here we focus on the decision tree. Recall that both bagging and random forest create multiple copies of the original training data using the bootstrap, fitting a separate decision tree to each copy and combining all the results to create a single prediction. Boosting also creates different trees but the trees are grown sequentially and each tree is a weak learner. Any modeling technique with tuning parameters can produce a range of learners, from weak to strong. You can easily make a weak learner by restricting the depth of the tree. There are different types of boosting. Here we introduce two main types: adaptive boosting and stochastic gradient boosting.

11.7.1 Adaptive Boosting

Yoav Freund and Robert Schapire (Freund and Schapire, 1997) came up the AdaBoost.M1 algorithm. Consider a binary classification problem where the response variable has two categories $Y \in \{-1, 1\}$. Given predictor matrix, X, construct a classifier $G(X)$ that predicts 1 or -1. The corresponding error rate in the training set is:

$$\bar{err} = \frac{1}{N}\Sigma_{i=1}^{N}I(y_i \neq G(x_i))$$

The algorithm produces a series of classifiers $G_m(x)$, $m = 1, 2, ..., M$ from different iterations. In each iteration, it finds the best classifier based on the current weights. The misclassified samples in the mth iteration will have higher weights in the $(m+1)^{th}$ iteration and the correctly classified samples will have lower weights. As it moves on, the algorithm will put more effort into the "difficult" samples until it can correctly classify them. So it requires the algorithm to change focus at each iteration. At each iteration, the algorithm will calculate a stage weight based on the error rate. The final prediction is a weighted average of all those weak classifiers using stage weights from all the iterations:

$$G(x) = sign(\Sigma_{m=1}^{M}\alpha_m G_m(x))$$

where $\alpha_1, \alpha_2, ..., \alpha_M$ are the weights from different iterations.

Algorithm 5 AdaBoost.M1

1: Response variables have two values: $+1$ and -1
2: Initialize the observation to have the same weights: $w_i = \frac{1}{N}, i = 1, ..., N$
3: **for** m = 1 to M **do**
4:     Fit a classifier $G_m(x)$ using weights w_i
5:     Compute the error rate: $err_m = \frac{\sum_{i=1}^{N} w_i I(y_i \neq G_m(x_i))}{\sum_{i=1}^{N} w_i}$
6:     Compute the stage weight: $\alpha_m = log\frac{1-err_m}{err_m}$
7:     Update $w_i = w_i \cdot exp[\alpha_m \cdot I(y_i \neq G_m(x_i))]$, $i = 1, 2, ..., N$
8: **end for**
9: Calculate the prediction: $G(x) = sign[\sum_{m=1}^{M} \alpha_m G_m(x)]$, where $sign(\cdot)$ means if $\cdot$ is positive, then the sample is classified as $+1$, -1 otherwise.

Since the classifier $G_m(x)$ returns discrete value, the AdaBoost.M1 algorithm is known as "Discrete AdaBoost" (Friedman et al., 2000). You can revise the above algorithm if it returns continuousf value, for example, a probability (Friedman et al., 2000). As mentioned before, boosting is a general approach that can be applied to different learners. Since you can easily create weak learners by limiting the depth of the tree, the boosting tree is a common method. Since the classification tree is a low bias/high variance technique, ensemble decreases model variance and lead to low bias/low variance model. See Breinman (Breiman, 1998) for more explanation about why the boosting tree performs well in general. However, boosting cannot significantly improve the low variance model. So applying boosting to K-Nearest Neighbor (KNN) doesn't lead to as good improvement as applying boosting to statistical learning methods like naive Bayes (Bauer and Kohavi, 1999).

11.7.2 Stochastic Gradient Boosting

As mentioned before, Friedman (Friedman et al., 2000) provided a statistical framework for the AdaBoost algorithm and pointed out that boosting can be considered as a forward stagewise additive model that minimizes exponential loss. The framework led to some generalized algorithms such as Real AdaBoost, Gentle AdaBoost, and LogitBoost. Those algorithms later were unified under a framework called gradient boosting machine. The last section of the chapter illustrates how boosting can be considered as an additive model.

Consider a 2-class classification problem. You have the response $y \in \{-1, 1\}$ and the sample proportion of class 1 from the training set is p. $f(x)$ is the model prediction in the range of $[-\infty, +\infty]$ and the predicted event probability is $\hat{p} = \frac{1}{1 + exp[-f(x)]}$. The gradient boosting for this problem is as follows:

Algorithm 6 Stochastic gradient boosting for 2-class classification

---

1: Response variables have two values: $+1$ and -1
2: Initialize all predictions to the sample log-odds: $f_i = log\frac{\hat{p}}{1-\hat{p}}$
3: **for** j=1 ... M **do**
4:     Compute predicted event probability: $\hat{p}_i = \frac{1}{1 + exp[-f_i(x)]}$
5:     Compute the residual (i.e. gradient): $z_i = y_i - \hat{p}_i$
6:     Randomly sample the training data
7:     Train a tree model on the random subset using the residuals
    as the outcome
8:     Compute the terminal node estimates of the Pearson resid-
    uals: $r_i = \frac{1/n\Sigma_i^n(y_i-\hat{p}_i)}{1/n\Sigma_i^n\hat{p}_i(1-\hat{p}_i)}$
9:     Update f: $f_i = f_i + \lambda f_i^{(j)}$
10: **end for**

When using the tree as the base learner, basic gradient boosting has two tuning parameters: tree depth and the number of iterations. You can further customize the algorithm by selecting a different loss function and gradient (Hastie et al., 2008). The final line of the loop includes a regularization strategy. Instead of adding $f_i^{(j)}$ to the previous iteration's f_i, only a fraction of the value is added. This fraction is called learning rate which is λ in the algorithm. It can take values between 0 and 1 which is another tuning parameter of the model.

The way to calculate variable importance in boosting is similar to a bagging model. You get variable importance by combining measures of importance across the ensemble. For example, we can calculate the Gini index improvement for each variable across all trees and use the average as the measurement of the importance.

Boosting is a very popular method for classification. It is one of the methods that can be directly applied to the data without requiring a great deal of time-consuming data preprocessing. Applying boosting on tree models significantly improves predictive accuracy. Some advantages of trees that are sacrificed by boosting are speed and interpretability.

Let's look at the R implementation.

```r
gbmGrid <- expand.grid(interaction.depth = c(1, 3, 5, 7, 9),
                       n.trees = 1:5,
                       shrinkage = c(.01, .1),
                       n.minobsinnode = c(1:10))

set.seed(100)
gbmTune <- caret::train(x = trainx,
                        y = trainy,
                        method = "gbm",
                        tuneGrid = gbmGrid,
                        metric = "ROC",
                        verbose = FALSE,
                        trControl = trainControl(method = "cv",
                                classProbs = TRUE,
                                savePredictions = TRUE))
```

```r
# only show part of the output
gbmTune
```

```
Stochastic Gradient Boosting

1000 samples
  11 predictor
   2 classes: 'Female', 'Male'

No pre-processing
Resampling: Cross-Validated (10 fold)
Summary of sample sizes: 899, 900, 900, 899, 899, 901, ...
Resampling results across tuning parameters:
```

shrinkage	interaction.depth	n.minobsinnode	n.trees	ROC	Sens	Spec
0.01	1	1	1	0.6821	1.00	0.00
0.01	1	1	2	0.6882	1.00	0.00
.						
.						
.						
0.01	5	8	4	0.7096	1.00	0.00
0.01	5	8	5	0.7100	1.00	0.00
0.01	5	9	1	0.7006	1.00	0.00
0.01	5	9	2	0.7055	1.00	0.00

```
 [ reached getOption("max.print") -- omitted 358 rows ]

ROC was used to select the optimal model using the largest value.
The final values used for the model were n.trees = 4,
interaction.depth = 3, shrinkage = 0.01 and n.minobsinnode = 6.
```

The results show that the tuning parameter settings that lead to the best ROC are n.trees = 4 (number of trees), interaction.depth = 3 (depth of tree), shrinkage = 0.01 (learning rate) and n.minobsinnode = 6 (minimum number of observations in each node).

Now, let's compare the results from the three tree models.

```
treebagRoc <- pROC::roc(response = bagTune$pred$obs,
                        predictor = bagTune$pred$Female,
```

```
                         levels = rev(levels(bagTune$pred$obs)))

rfRoc <- pROC::roc(response = rfTune$pred$obs,
              predictor = rfTune$pred$Female,
              levels = rev(levels(rfTune$pred$obs)))

gbmRoc <- pROC::roc(response = gbmTune$pred$obs,
               predictor = gbmTune$pred$Female,
               levels = rev(levels(gbmTune$pred$obs)))

plot.roc(rpartRoc,
    type = "s",
    print.thres = c(.5), print.thres.pch = 16,
    print.thres.pattern = "", print.thres.cex = 1.2,
    col = "black", legacy.axes = TRUE,
    print.thres.col = "black")

plot.roc(treebagRoc,
    type = "s",
    add = TRUE,
    print.thres = c(.5), print.thres.pch = 3,
    legacy.axes = TRUE, print.thres.pattern = "",
    print.thres.cex = 1.2,
    col = "red", print.thres.col = "red")

plot.roc(rfRoc,
    type = "s",
    add = TRUE,
    print.thres = c(.5), print.thres.pch = 1,
    legacy.axes = TRUE, print.thres.pattern = "",
    print.thres.cex = 1.2,
    col = "green", print.thres.col = "green")

plot.roc(gbmRoc,
    type = "s",
    add = TRUE,
```

```
    print.thres = c(.5), print.thres.pch = 10,
    legacy.axes = TRUE, print.thres.pattern = "",
    print.thres.cex = 1.2,
    col = "blue", print.thres.col = "blue")

legend(0.2, 0.5, cex = 0.8,
        c("Single Tree", "Bagged Tree",
          "Random Forest", "Boosted Tree"),
        lwd = c(1, 1, 1, 1),
        col = c("black", "red", "green", "blue"),
        pch = c(16, 3, 1, 10))
```

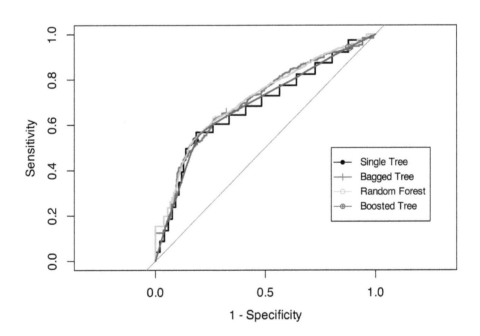

Since the data here doesn't have many variables, we don't see
a significant difference among the models. But you can still see
those ensemble methods are better than a single tree. In most
of the real applications, ensemble methods perform much better.
Random forest and boosting trees can be a baseline model. Before
exploring different models, you can quickly run a random forest to
see the performance and then try to improve that performance. If

the performance you got from the random forest is not too much better than guessing, you should consider collecting more data or reviewing the problem to frame it a different way instead of trying other models. Because it usually means the current data is not enough to solve the problem.

12

Deep Learning

Deep learning has become widely known in recent years due to its applications in language, voice, image, and self-driving cars. However, its roots can be traced back to the 1940s. For example, the binary perceptron classifier, invented in the late 1950s, uses a linear combination of input signals and a step activation function. This is the same as a single neuron in a modern deep learning network that uses the same linear combination of input signals from neurons at the previous layer and a more efficient nonlinear activation function.

The perceptron model was further defined by minimizing the classification error and trained by using one data point at a time to update the model parameters during the optimization process. Modern neural networks are trained similarly by minimizing a loss function but with more modern optimization algorithms such as stochastic gradient descent and its variations.

Even though the theoretical foundation of deep learning has been continually developed in the past few decades, real-world applications of deep learning are fairly recent due to some real world constraints: data, network structure, algorithm, and computation power.

Data

We are all familiar with all sorts of data today: structured tabulated data in database tables or CSV files, freeform text, images, and other unstructured datasets. However, historical datasets tend to be relatively small in size, particularly when it comes to data with accurately labeled ground truth. Statisticians have been working on datasets with only a few thousand rows and a few dozen columns for decades to solve business problems. Even with modern computers,

DOI: 10.1201/9781351132916-12

the size of the data is usually limited to the memory of a computer. Now we understand that deep learning applications require a much larger dataset than traditional machine learning methods. Generally, this dataset must consist of millions of samples with accurate labels for supervised deep learning models.

The first widely used large dataset with accurate labels was the ImageNet dataset which was created in 2010. It now contains more than 14 million images with more than 20k synsets (i.e. meaningful categories). Every image in the dataset was human-annotated with quality control to ensure accurate ground truth labels.

The Large Scale Visual Recognition Challenge (ILSVRC) was a direct result of ImageNet, which evaluated different algorithms in image-related tasks. It provided a perfect stage for deep learning applications to debut to the general public.

In 2010 and 2011, the best record of error from traditional image classification methods was around 26%. In 2012, a method based on the convolutional neural network achieved a state-of-the-art error rate of around 16%, a dramatic improvement from the traditional methods.

The internet has made it possible for a huge amount of text, voice, image, and video data to be created and stored. This has enabled the use of deep learning for applications such as image classification, voice recognition, and natural language understanding. Data is the fuel for deep learning engines. With more and more varieties of data created, captured, and saved, new applications of deep learning are being discovered every day.

Network Structure

Lacking high-quality high-volume data was not the only constraint in the early years of deep learning. For perceptron with one single neuron, it is simply a linear classifier. However, real applications are nearly always non-linear. To solve this problem, we have to expand one neuron to multiple layers with multiple neurons per layer. This multi layer perceptron is also referred to as a feedforward neural network.

In the 1990s, the universal approximation theorem was proven, assuring us that a feedforward network with a single hidden layer containing a finite number of neurons can approximate continuous functions. Even though the one-layer neural network theoretically can solve a general non-linear problem, the reality is that we have grown the neural network to many layers of neurons. The number of layers in the network is the "depth" of a network. Loosely speaking, deep learning is a neural network with many layers (i.e. the depth is deep).

The **Multilayer Perceptron (MLP)** is the basic structure for the modern deep learning applications. It can be used for classification and regression problems, where the response variables is the output and a collection of explanatory variables is the input (i.e. the traditionally structured datasets). Many of the problems that can be solved using classical classification methods, such as random forest can be solved by MLP.

However, MLP is not the best option for image and language-related tasks. For image-related tasks, pixels from a local neighbor region collectively provide useful information to solve a task. To take advantage of the 2D spatial relationship among pixels, the **Convolutional Neural Network (CNN)** structure is a better choice. For language-related tasks, the sequence of the text provides additional information than just a collection of single words. The **Recurrent Neural Network (RNN)** is a better structure for such sequence-related data.

There are other more complicated neural network structures and it is still a fast-developing area. MLP, CNN, and RNN are just the starting point of deep learning methods.

Algorithm

In addition to data and neural network structure, there were a few key algorithm breakthroughs that enabled the widespread adoption of deep learning. For an entry-level neural network structure, there are hundreds of thousands of parameters to be estimated from the data.

With a large amount of training data, stochastic gradience decent and mini-batch gradience decent are efficient ways to utilize a subset of training data to update the model parameters. Within the process, one of the key steps is back-propagation, which was introduced in the 1980s for efficient weight update.

There is a non-linear activation for each neuron in deep learning models, and **sigmoid** or **hyperbolic tangent** functions were often used. However, it had the problem of gradient vanishing when the number of layers of the network grew large (i.e. deeper network). To solve this problem, the **rectified linear unit (ReLu)** was introduced to deep learning in the 2000s and it increased the convergence speed dramatically. ReLu is simple (i.e. y = x when x >= 0 and y = 0 otherwise), but it cleverly solved one of the big headaches in deep learning. We will discuss activation functions in more detail in section 12.1.4.

With hundreds of thousands of parameters, deep learning is prone to overfitting. To address this, dropout, a form of regularization, was introduced in 2012. It randomly drops out a certain percentage of neurons during the optimization process, making the model more robust. This is similar to the concept of random forest, where features and training data are randomly chosen.

There are many other algorithm improvements to get better models, such as batch normalization and using residuals from previous layers. With backpropagation in stochastic gradience decent, ReLu activation function, dropout, and other techniques, modern deep learning methods have begun to outperform traditional machine learning methods.

Computation Power

Data, network structure, and algorithms are ready for modern deep learning, but it still requires a certain amount of computation power for training. This framework involves heavy linear algebra operations with large matrices and tensors. These types of operations are much faster on modern graphical processing units (GPUs) than the computer's central processing units (CPU).

With the vast potential applications of deep learning, major tech companies have heavily contributed to open-source deep learning frameworks. For example, Google has open-sourced its TensorFlow framework, Facebook has open-sourced its PyTorch framework, and Amazon has significantly contributed to the MXNet open-source framework. Thousands of software developers and scientists are behind these deep learning frameworks, allowing users to confidently pick one framework and start training their deep learning models in popular cloud environments. Much of the heavy lifting to train a deep learning model has been embedded in these open-source frameworks, and there are also many pre-trained models available for users to adopt.

Now, users can enjoy the relatively easy access to software and hardware to develop their own deep learning applications. In this book, we will demonstrate deep learning examples using Keras, a high-level abstraction of TensorFlow.

In summary, deep learning has not just developed in the past few years, but has been the subject of research for decades. The accumulation of data, the development of new optimization algorithms, and the improvement of computation power have enabled every day deep learning applications. In the foreseeable future, deep learning will continue to revolutionize machine learning methods across many more areas.

12.1 Feedforward Neural Network

Feedforward Neural Networks (FFNNs) are a type of neural network where information flows in a single direction, from input to output. In an FFNN, the input data is processed in a series of layers, with each layer connected to the next. The output of each layer is calculated based on the inputs from the previous layer. The outputs from the final layer are the results of the network.

12.1.1 Logistic Regression as Neural Network

Before delving into deep neural networks, let's take a look at a statistical model that many people are familiar with: logistic regression. It can be viewed as a 0-hidden layer FFNN. For a binary classification problem, such as a spam classifier, given m samples $\{(x^{(1)}, y^{(1)}), (x^{(2)}, y^{(2)}), ..., (x^{(m)}, y^{(m)})\}$, we need to use the input feature $x^{(i)}$ (they may be the frequency of various words such as "money," special characters like dollar signs, and the use of capital letters in the message etc.) to predict the output $y^{(i)}$ (if it is a spam email). Assume that for each sample i, there are n_x input features. Then we have:

$$X = \begin{bmatrix} x_1^{(1)} & x_1^{(2)} & \cdots & x_1^{(m)} \\ x_2^{(1)} & x_2^{(2)} & \cdots & x_2^{(m)} \\ \vdots & \vdots & \vdots & \vdots \\ x_{n_x}^{(1)} & x_{n_x}^{(2)} & \cdots & x_{n_x}^{(m)} \end{bmatrix} \in \mathbb{R}^{n_x \times m} \qquad (12.1)$$

$$y = [y^{(1)}, y^{(2)}, \dots, y^{(m)}] \in \mathbb{R}^{1 \times m}$$

To predict if sample i is a spam email, we first get the inactivated **neuro** $z^{(i)}$ by a linear transformation of the input $x^{(i)}$, which is $z^{(i)} = w^T x^{(i)} + b$. Then we apply a function to "activate" the neuro $z^{(i)}$ and we call it "activation function." In logistic regression, the activation function is sigmoid function and the "activated" $z^{(i)}$ is the prediction:

$$\hat{y}^{(i)} = \sigma(w^T x^{(i)} + b)$$

where $\sigma(z) = \frac{1}{1+e^{-z}}$. The following figure summarizes the process:

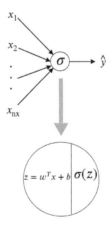

There are two types of layers. The last layer connects directly to the output. All the rest are *intermediate layers*. Depending on your definition, we call it "0-layer neural network" where the layer count only considers *intermediate layers*. To train the model, you need a cost function which is defined as equation (12.2).

$$J(w,b) = \frac{1}{m}\Sigma_{i=1}^{m} L(\hat{y}^{(i)}, y^{(i)}) \qquad (12.2)$$

where

$$L(\hat{y}^{(i)}, y^{(i)}) = -y^{(i)}log(\hat{y}^{(i)}) - (1 - y^{(i)})log(1 - \hat{y}^{(i)})$$

To fit the model is to minimize the cost function.

12.1.2 Stochastic Gradient Descent

The general approach to minimize $J(w,b)$ is by gradient descent, also known as *back-propagation*. The optimization process is a forward and backward sweep over the network.

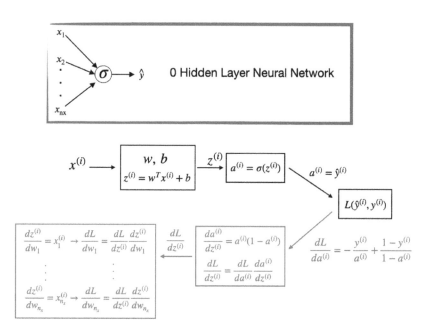

The forward propagation takes the current weights, calculates the prediction and cost. The backward propagation computes the gradient descent for the parameters by the chain rule of differentiation. In logistic regression, it is straightforward to calculate the gradient with respect to the parameters (w, b).

Let's look at the Stochastic Gradient Descent (SGD) for logistic regression across m samples. SGD updates one sample each time. The detailed process is as follows.

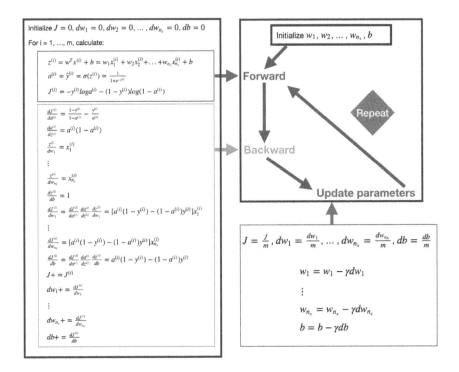

First initialize w_1, w_2, ... , w_{n_x}, and b. Then plug in the initialized value to the forward and backward propagation. The forward propagation takes the current weights and calculates the prediction $\hat{h}^{(i)}$ and cost $J^{(i)}$. The backward propagation calculates the gradient descent for the parameters. After iterating through all m samples, you can calculate gradient descent for the parameters. Then update the parameter by:

$$w := w - \gamma \frac{\partial J}{\partial w}$$
$$b := b - \gamma \frac{\partial J}{\partial b}$$

Repeat the forward and backward process using the updated parameter until the cost J stabilizes.

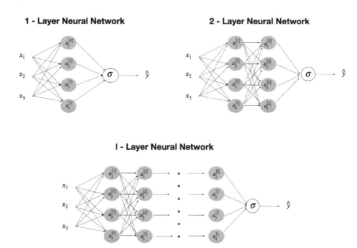

FIGURE 12.1

Feedforward neural network

12.1.3 Deep Neural Network

Before people coined the term *"deep learning,"* a neural network referred to a *single hidden layer network*. Neural networks with more than one layers are called *deep learning*. Network with the structure in figure 12.1 is the **multiple layer perceptron (MLP) or feedforward neural network (FFNN)**.

Let's look at a simple one-hidden-layer neural network (figure 12.2). First only consider one sample. From left to right, there is an input layer with 3 features (x_1, x_2, x_3), a hidden layer with four neurons and an output later to produce a prediction $\hat{y}$.

From input to the first hidden layer

Each inactivated neuron on the first layer is a linear transformation of the input vector x. For example, $z_1^{[1]} = w_1^{[1]T} x^{(i)} + b_1^{[1]}$ is the first inactivated neuron for hidden layer one. **We use superscript [l] to denote a quantity associated with the lth layer and the subscript i to denote the ith entry of a vector (a neuron or feature).** Here $w^{[1]}$ and $b_1^{[1]}$ are the weight and bias parameters for layer 1. $w^{[1]}$ is a 4×1 vector and hence $w_1^{[1]T} x^{(i)}$ is a linear

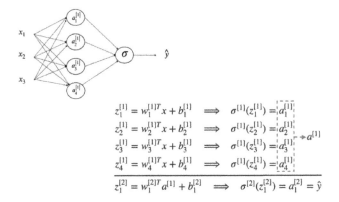

$$z_1^{[1]} = w_1^{[1]T}x + b_1^{[1]} \implies \sigma^{[1]}(z_1^{[1]}) = a_1^{[1]}$$
$$z_2^{[1]} = w_2^{[1]T}x + b_2^{[1]} \implies \sigma^{[1]}(z_2^{[1]}) = a_2^{[1]}$$
$$z_3^{[1]} = w_3^{[1]T}x + b_3^{[1]} \implies \sigma^{[1]}(z_3^{[1]}) = a_3^{[1]} \quad \to a^{[1]}$$
$$z_4^{[1]} = w_4^{[1]T}x + b_4^{[1]} \implies \sigma^{[1]}(z_4^{[1]}) = a_4^{[1]}$$
$$z_1^{[2]} = w_1^{[2]T}a^{[1]} + b_1^{[2]} \implies \sigma^{[2]}(z_1^{[2]}) = a_1^{[2]} = \hat{y}$$

FIGURE 12.2
1-Layer neural network

combination of the four input features. Then use a sigmoid function $\sigma(\cdot)$ to activate the neuron $z_1^{[1]}$ to get $a_1^{[1]}$.

From the first hidden layer to the output

Next, do a linear combination of the activated neurons from the first layer to get inactivated output, $z_1^{[2]}$. And then activate the neuron to get the predicted output $\hat{y}$. The parameters to estimate in this step are $w^{[2]}$ and $b_1^{[2]}$.

If you fully write out the process, it is the bottom right of figure 12.2. When you implement a neural network, you need to do similar calculation four times to get the activated neurons in the first hidden layer. Doing this with a `for` loop is inefficient. So people vectorize the four equations. Take an input and compute the corresponding z and a as a vector. You can vectorize each step and get the following representation:

$$z^{[1]} = W^{[1]}x + b^{[1]} \qquad \sigma^{[1]}(z^{[1]}) = a^{[1]}$$
$$z^{[2]} = W^{[2]}a^{[1]} + b^{[2]} \qquad \sigma^{[2]}(z^{[2]}) = a^{[2]} = \hat{y}$$

$b^{[1]}$ is the column vector of the four bias parameters shown above. $z^{[1]}$ is a column vector of the four non-active neurons. When you apply an activation function to a matrix or vector, you apply it element-wise. $W^{[1]}$ is the matrix by stacking the four row-vectors:

$$W^{[1]} = \begin{bmatrix} w_1^{[1]T} \\ w_2^{[1]T} \\ w_3^{[1]T} \\ w_4^{[1]T} \end{bmatrix}$$

So if you have one sample, you can go through the above forward propagation process to calculate the output $\hat{y}$ for that sample. If you have m training samples, you need to repeat this process each of the m samples. **We use superscript (i) to denote a quantity associated with** i^{th} **sample.** You need to do the same calculation for all m samples.

For i $= 1$ to m, do:

$$z^{[1](i)} = W^{[1]}x^{(i)} + b^{[1]} \qquad \sigma^{[1]}(z^{[1](i)}) = a^{[1](i)}$$
$$z^{[2](i)} = W^{[2]}a^{[1](i)} + b^{[2]} \qquad \sigma^{[2]}(z^{[2](i)}) = a^{[2](i)} = \hat{y}^{(i)}$$

Recall that we defined the matrix X to be equal to our training samples stacked up as column vectors in equation (12.1). We do a similar thing here to stack vectors with the superscript (i) together across m samples. This way, the neural network computes the outputs on all the samples on at the same time:

$$Z^{[1]} = W^{[1]}X + b^{[1]} \qquad \sigma^{[1]}(Z^{[1]}) = A^{[1]}$$
$$Z^{[2]} = W^{[2]}A^{[1]} + b^{[2]} \qquad \sigma^{[2]}(Z^{[2]}) = A^{[2]} = \hat{Y}$$

where

$$X = \begin{bmatrix} | & | & & | \\ x^{(1)} & x^{(1)} & \cdots & x^{(m)} \\ | & | & & | \end{bmatrix},$$

$$A^{[l]} = \begin{bmatrix} | & | & & | \\ a^{[l](1)} & a^{[l](1)} & \cdots & a^{[l](m)} \\ | & | & & | \end{bmatrix}_{l=1 \ or \ 2},$$

$$Z^{[l]} = \begin{bmatrix} | & | & & | \\ z^{[l](1)} & z^{[l](1)} & \cdots & z^{[l](m)} \\ | & | & & | \end{bmatrix}_{l=1 \ or \ 2}$$

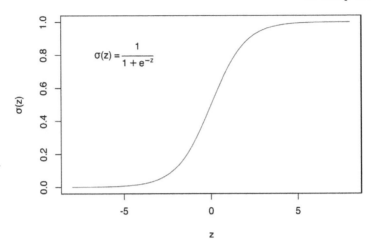

FIGURE 12.3
Sigmoid function

You can add layers like this to get a deeper neural network as shown in the bottom right of figure 12.1.

12.1.4 Activation Function

- Sigmoid and Softmax Function

We have used the sigmoid (or logistic) activation function. The function is S-shape with an output value between 0 to 1. Therefore it is used as the output layer activation function to predict the probability **when the response y is binary**. However, it is rarely used as an intermediate layer activation function. One of the main reasons is that when z is away from 0, then the derivative of the function drops fast which slows down the optimization process through gradient descent. Even with the fact that it is differentiable provides some convenience, the decreasing slope can cause a neural network to get stuck at the training time.

When the output has more than 2 categories, people use softmax function as the output layer activation function.

$$f_i(\mathbf{z}) = \frac{e^{z_i}}{\sum_{j=1}^{J} e^{z_j}} \tag{12.3}$$

where $\mathbf{z}$ is a vector.

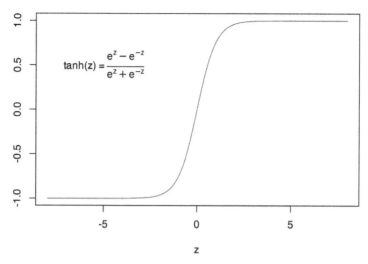

FIGURE 12.4
Hyperbolic tangent function

- Hyperbolic Tangent Function (tanh)

Another activation function with a similar S-shape is the hyperbolic tangent function. It often works better than the sigmoid function as the intermediate layer.

$$tanh(z) = \frac{e^z - e^{-z}}{e^z + e^{-z}} \qquad (12.4)$$

The tanh function crosses point (0, 0) and the value of the function is between 1 and -1 which makes the mean of the activated neurons closer to 0. The sigmoid function doesn't have that property. When you preprocess the training input data, you sometimes center the data so that the mean is 0. The tanh function is doing that data processing in some way which makes learning for the next layer a little easier. This activation function is used a lot in the recurrent neural networks where you want to polarize the results.

- Rectified Linear Unit (ReLU) Function

The most popular activation function is the Rectified Linear Unit (ReLU) function. It is a piecewise function, or a half rectified function:

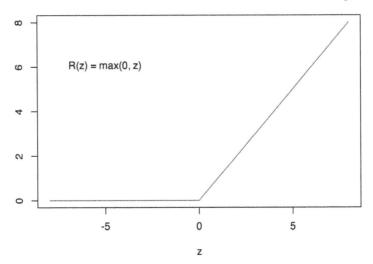

FIGURE 12.5
Rectified linear unit function

$$R(z) = max(0, z) \tag{12.5}$$

The derivative is 1 when z is positive and 0 when z is negative. You can define the derivative as either 0 or 1 when z is 0.

The advantage of the ReLU is that when z is positive, the derivative doesn't vanish as z getting larger. So it leads to faster computation than sigmoid or tanh. It is non-linear with an unconstrained response. However, the disadvantage is that when z is negative, the derivative is 0. It may not map the negative values appropriately. In practice, this doesn't cause too much trouble but there is another version of ReLu called leaky ReLu that attempts to solve the dying ReLU problem. The leaky ReLu is

$$R(z)_{Leaky} = \begin{cases} z & z \geq 0 \\ az & z < 0 \end{cases}$$

Instead of being 0 when z is negative, it adds a slight slope such as $a = 0.01$ as shown in figure 12.6.

You may notice that all activation functions are non-linear. Since the composition of two linear functions is still linear, using a linear activation function doesn't help to capture more information. That is why you don't see people use a linear activation function

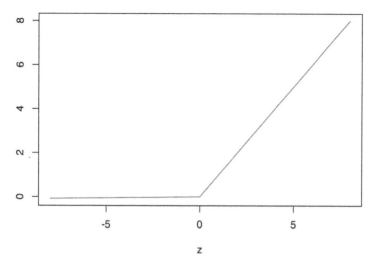

FIGURE 12.6
Leaky rectified linear unit function

in the intermediate layer. One exception is when the output y is continuous, you may use linear activation function at the output layer. To sum up, for intermediate layers:

- ReLU is usually a good choice. If you don't know what to choose, then start with ReLU. Leaky ReLu usually works better than the ReLU but it is not used as much in practice. Either one works fine. Also, people usually use a = 0.01 as the slope for leaky ReLU. You can try different parameters but most of the people a = 0.01.
- tanh is used sometimes especially in recurrent neural network. But you nearly never see people use sigmoid function as intermediate layer activation function.

For the output layer:

- When it is binary classification, use sigmoid with binary cross-entropy as loss function.
- When there are multiple classes, use softmax function with categorical cross-entropy as loss function.
- When the response is continuous, use identity function (i.e. y = x).

12.1.5   Optimization

So far, we have introduced the core components of deep learning: architecture, layer, weight, activation function, and loss function. Now
that the architecture is set, we need to determine how to update
the network based on a loss function (also known as an objective
function). In this section, we will explore different optimization
algorithms that can speed up the training process.

12.1.5.1   Batch, Mini-batch, Stochastic Gradient Descent

Stochastic Gradient Descent (SGD) updates model parameters
one sample at a time. We demonstrated SGD for logistic regression
across m samples in section 12.1.1. If you process the entire training
set each time to calculate the gradient, known as **Batch Gradient
Descent (BGD)**. The vector representation (section 12.1.3) using
all m is an example of BGD.

In deep learning applications, the training set is often huge, containing hundreds of thousands or millions of samples. If processing
all the samples only leads to one step of gradient descent, it can be
slow. To improve the algorithm, we can split the training set into
smaller subsets, known as mini-batches, and fit the model using on
subset at a time. **Mini-batch Gradient Descent (MGD)** is to
split the training set to be smaller mini-batches.

For example, if the mini-batch size is 1000, the algorithm will
process 1000 samples each time, calculate the gradient and update
the parameters. It then moves on to the next mini-batch set until
it goes through the entire training set. This is known as one pass
through training set using mini-batch gradient descent or one epoch.

Epoch is a common keyword in deep learning, which means a
single pass through the training set. For example, if the training
set has 60,000 samples, one epoch leads to 60 gradient descent
steps. Then, the process starts over and takes another pass through
the training set. It means one more decision to make, the optimal
number of epochs. We decide how many epochs to use by looking
at the trends of performance metrics on a holdout set of training
data. We discussed the data splitting and sampling in section 7.2.
People often use a single holdout set to tune the model in deep
learning. It is important, however, to use a large enough holdout

set to gain confidence in the model's overall performance. After that, you can evaluate the chosen model on your test set, which was used during the training process. MGD is the preferred method when training on a large data set.

$$x = \quad [\underbrace{x^{(1)}, x^{(2)}, \cdots, x^{(1000)}}_{mini-batch\ 1} / \ \cdots / \cdots x^{(m)}]$$
$$(n_x, m)$$

$$y = \quad [\underbrace{y^{(1)}, y^{(2)}, \cdots, y^{(1000)}}_{mini-batch\ 1} / \ \cdots / \cdots y^{(m)}]$$
$$(1, m)$$

- Mini-batch size = m: batch gradient descent takes too long per iteration.
- Mini-batch size = 1: stochastic gradient descent loses speed from vectorization.
- Mini-batch size in between: mini-batch gradient descent makes progress without processing the entire training set. The typical batch sizes are $2^6 = 64$, $2^7 = 128$, $2^8 = 256$, $2^9 = 512$.

12.1.5.2 Optimization Algorithms

Throughout the history of deep learning, researchers have proposed various optimization algorithms and showed that they worked well in a specific scenario. But the optimization algorithms didn't generalize well across a wide range of neural networks. Therefore, you must experiment with different optimizers for your application. We will introduce three commonly used optimizers here, all of which are based on the concept of exponentially weighted averages. To gain an understanding of this concept, let's look at a hypothetical example shown in figure 12.7.

We have two parameters, b and w. The blue dot represents the current parameter value, while the red point is the optimum value we want to reach. The current value is close to the target vertically, but far away horizontally. In this situation, we hope that learning on b will be slower and learning on w will be faster. To adjust this, we can use the average of the gradients from different iterations instead of the current iteration to update the parameter.

Vertically, since the current value is close to the target value of parameter b, the gradient of b is likely to jump between positive

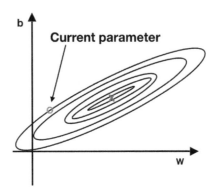

FIGURE 12.7

The intuition behind the weighted average gradient

and negative values. The average tends to cancel out the positive and negative derivatives along the vertical direction and slow down the oscillations.

Horizontally, since it is still further away from the target value of parameter w, all the gradients are likely pointing in the same direction. Using an average won't have too much impact there. That is the fundamental idea behind many different optimizers: adjust the learning rate using a weighted average of various iterations' gradients.

Exponentially Weighted Averages

Before diving into more complex optimization algorithms that utilize this concept, let's first cover the basics of weighted moving average.

Suppose we have the following 100 days' temperature data:

$$\theta_1 = 49F, \theta_2 = 53F, \dots, \theta_{99} = 70F, \theta_{100} = 69F$$

The weighted average is defined as:

$$V_t = \beta V_{t-1} + (1 - \beta)\theta_t$$

And we have:

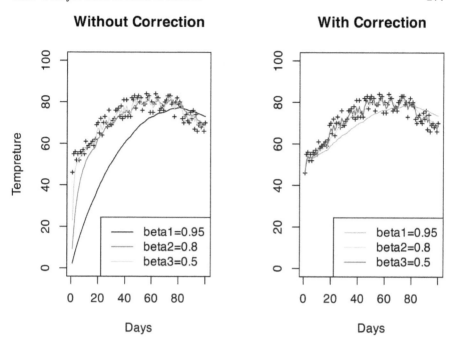

FIGURE 12.8
Exponentially weighted averages with and without corrrection

$$V_0 = 0$$
$$V_1 = \beta V_1 + (1 - \beta)\theta_1$$
$$V_2 = \beta V_1 + (1 - \beta)\theta_2$$
$$\vdots$$
$$V_{100} = \beta V_{99} + (1 - \beta)\theta_{100}$$

For example, for $\beta = 0.95$:

$$V_0 = 0$$
$$V_1 = 0.05\theta_1 \qquad \ldots\ldots$$
$$V_2 = 0.0475\theta_1 + 0.05\theta_2$$

The black line in the left plot of figure 12.8 is the exponentially weighted averages of simulated temperature data with $\beta = 0.95$. V_t is approximately average over the previous $\frac{1}{1-\beta}$ days. So $\beta = 0.95$ approximates a 20 days' average. The red line corresponds to $\beta = 0.8$, which approximates 5 days' average. As β increases, it averages over a larger window of the previous values, and hence the curve gets smoother. A larger β also means that it gives the current value θ_t less weight $(1 - \beta)$, and the average adapts more slowly. It is easy to see from the plot that the averages at the beginning are more biased. The bias correction can help to achieve a better estimate:

$$V_t^{corrected} = \frac{V_t}{1 - \beta^t}$$

$$V_1^{corrected} = \frac{V_1}{1 - 0.95} = \theta_1$$

$$V_2^{corrected} = \frac{V_2}{1 - 0.95^2} = 0.4872\theta_1 + 0.5128\theta_2$$

For $\beta = 0.95$, the origional $V_2 = 0.0475\theta_1 + 0.05\theta_2$ which is a small fraction of both θ_1 and θ_2. That is why it starts so much lower with big bias. After correction, $V_2^{corrected} = 0.4872\theta_1 + 0.5128\theta_2$ is a weighted average with two weights added up to 1 which revmoves the bias. Notice that as t increases, β^t converges to 0 and $V^{corrected}$ converges to V^t.

```
# Define a, b and c
a = -30/3479
b = -120 * a
c = 3600 * a + 80


# Generate temperature data for 100 days
day = c(1:100)
theta = a * day^2 + b * day + c + runif(length(day), -5, 5)
theta = round(theta, 0)


par(mfrow=c(1,2))
plot(day, theta, cex = 0.5, pch = 3, ylim = c(0, 100),
     main = "Without Correction",
     xlab = "Days", ylab = "Tempreture")


beta1 = 0.95
beta2 = 0.8
beta3 = 0.5
```

```r
exp_weight_avg = function(beta, theta) {
  v = rep(0, length(theta))

  for (i in 1:length(theta)) {
    if (i == 1) {
      v[i] = (1 - beta) * theta[i]
    } else {
      v[i] = beta * v[i - 1] + (1 - beta) * theta[i]
    }
  }
  return(v)
}


v1 = exp_weight_avg(beta = beta1, theta)
v2 = exp_weight_avg(beta = beta2, theta)
v3 = exp_weight_avg(beta = beta3, theta)


lines(day, v1, col = 1)
lines(day, v2, col = 2)
lines(day, v3, col = 3)
legend("bottomright",
       paste0(c("beta1=","beta2=","beta3="),
              c(beta1, beta2, beta3)), col = c(1:3),
       lty = 1)


exp_weight_avg_correct = function(beta, theta) {
  v = rep(0, length(theta))

  for (i in 1:length(theta)) {
    if (i == 1) {
      v[i] = (1 - beta) * theta[i]
    } else {
      v[i] = beta * v[i - 1] + (1 - beta) * theta[i]
    }
  }
  v = v/(1 - beta^c(1:length(v)))
  return(v)
```

```
}

v1_correct = exp_weight_avg_correct(beta = beta1, theta)
v2_correct = exp_weight_avg_correct(beta = beta2, theta)
v3_correct = exp_weight_avg_correct(beta = beta3, theta)

plot(day, theta, cex = 0.5, pch = 3, ylim = c(0,100),
     main = "With Correction",
     xlab = "Days", ylab = "")

lines(day, v1_correct, col = 4)
lines(day, v2_correct, col = 5)
lines(day, v3_correct, col = 6)
legend("bottomright",
       paste0(c("beta1=","beta2=","beta3="),
              c(beta1, beta2, beta3)),
       col = c(4:6), lty = 1)
```

The following code simulates a set of temperature data and applies exponentially weighted average with and without correction using various β values (0.5, 0.8, 0.95).

How do we apply the exponentially weighted average to optimization? Instead of using the gradients (dw and db) to update the parameters directly, we use the gradients' exponentially weighted average. There are various optimizers built on this idea. We will look at three of them: Momentum, Root Mean Square Propagation (RMSprop), and Adaptive Moment Estimation (Adam).

Momentum

The momentum algorithm uses the exponentially weighted average of gradients to update the parameters. On iteration t, compute dw, db using samples in one mini-batch and update the parameters as follows:

$$V_{dw} = \beta V_{dw} + (1 - \beta)dw$$

$$V_{db} = \beta V_{db} + (1 - \beta)db$$

$$w = w - \alpha V_{dw}; \quad b = b - \alpha V_{db}$$

The learning rate α and weighted average parameter β are hyperparameters. The most common and robust choice is $\beta = 0.9$. This algorithm does not use bias correction, as the average will warm up after a dozen iterations and no longer be biased. In general, the momentum algorithm works better than the original gradient descent without any average.

Root Mean Square Propagation (RMSprop)

The Root Mean Square Propagation (RMSprop) is another algorithm that applies the concept of exponentially weighted average. On iteration t, compute dw and db using the current mini-batch. Instead of V, it calculates the weighted average of the squared gradient descents. When dw and db are vectors, the squaring is an element-wise operation.

$$S_{dw} = \beta S_{dw} + (1 - \beta)dw^2$$

$$S_{db} = \beta S_{db} + (1 - \beta)db^2$$

Then, update the parameters as follows:

$$w = w - \alpha \frac{dw}{\sqrt{S_{dw}}}; \quad b = b - \alpha \frac{db}{\sqrt{S_{db}}}$$

The RMSprop algorithm divides the learning rate for a parameter by a weighted average of recent gradients' magnitudes for that parameter. The goal is still to adjust the learning speed. Recall the example that illustrates the intuition behind it. When parameter b is close to its target value, we want to decrease the oscillations along the vertical direction.

Adaptive Moment Estimation (Adam)

The Adaptive Moment Estimation (Adam) algorithm is, in some way, a combination of momentum and RMSprop. On iteration t, compute dw, db using the current mini-batch. Then calculate both V and S using the gradient descent.

$$\begin{cases} V_{dw} = \beta_1 V_{dw} + (1-\beta_1)dw \\ V_{db} = \beta_1 V_{db} + (1-\beta_1)db \end{cases} \quad momentum\ update\ \beta_1$$

$$\begin{cases} S_{dw} = \beta_2 S_{dw} + (1-\beta_2)dw^2 \\ S_{db} = \beta_2 S_{db} + (1-\beta_2)db^2 \end{cases} \quad RMSprop\ update\ \beta_2$$

The Adam algorithm implements bias correction.

$$\begin{cases} V_{dw}^{corrected} = \frac{V_{dw}}{1-\beta_1^t} \\ V_{db}^{corrected} = \frac{V_{db}}{1-\beta_1^t} \end{cases} ; \quad \begin{cases} S_{dw}^{corrected} = \frac{S_{dw}}{1-\beta_2^t} \\ S_{db}^{corrected} = \frac{S_{db}}{1-\beta_2^t} \end{cases}$$

And it updates the parameter using both corrected V and S, with a tiny positive number (ϵ) to make sure the denominator is larger than zero. The choice of ϵ doesn't matter much, and the inventors of the Adam algorithm recommended $\epsilon = 10^{-8}$. For hyperparameter β_1 and β_2, the common settings are $\beta_1 = 0.9$ and $\beta_2 = 0.999$.

$$w = w - \alpha \frac{V_{dw}^{corrected}}{\sqrt{S_{dw}^{corrected}} + \epsilon} ; \quad b = b - \alpha \frac{V_{db}^{corrected}}{\sqrt{S_{db}^{corrected}} + \epsilon}$$

12.1.6 Deal with Overfitting

The biggest problem for deep learning is overfitting. It happens when the model learns too much from the data. We discussed this in more detail in section 7.1. A common way to diagnose the problem is to use cross-validation (section 7.2). You can recognize the problem when the model fits well on the training data but gives poor predictions on the testing data. To prevent the model from over learning the data, one way is to limit model complexity. There are several approaches to that.

12.1.6.1 Regularization

For logistic regression, we can add a penalty term:

$$\min_{w,b} J(w,b) = \frac{1}{m}\Sigma_{i=1}^m L(\hat{y}^{(i)}, y^{(i)}) + penalty$$

Common penalties are L1 or L2 as follows:

$$L_2 \; penalty = \frac{\lambda}{2m} \parallel w \parallel_2^2 = \frac{\lambda}{2m} \Sigma_{i=1}^{n_x} w_i^2$$

$$L_1 \; penalty = \frac{\lambda}{m} \Sigma_{i=1}^{n_x} |w|$$

For neural network,

$$J(w^{[1]}, b^{[1]}, \ldots, w^{[L]}, b^{[L]}) = \frac{1}{m} \Sigma_{i=1}^m L(\hat{y}^{(i)}, y^{(i)}) + \frac{\lambda}{2} \Sigma_{l=1}^L \parallel w^{[l]} \parallel_F^2$$

where

$$\parallel w^{[l]} \parallel_F^2 = \Sigma_{i=1}^{n^{[l]}} \Sigma_{j=1}^{n^{[l-1]}} (w_{ij}^{[l]})^2$$

Many people call it "Frobenius Norm" instead of L2-norm.

12.1.6.2 Dropout

Dropout is another powerful regularization technique. In chapter 11, we mentioned that the random forest model de-correlates the trees by randomly selecting a subset of features. Dropout applies a similar concept during the parameter estimation process.

Dropout temporally freezes a randomly selected subset of nodes at a specific layer in the neural network during the optimization process to reduce overfitting. When applying dropout to a particular layer and mini-batch, a pre-set percentage (e.g. 30%) is removed from the layer's nodes. The output from the 30% removed nodes is set to zero. During backpropagation, the remaining parameters are updated for a much-diminished network. This random dropout must be repeated for each mini-batch.

To normalize the output with all nodes in this layer, the output must be scaled up accordingly to the same percentage to make sure the dropped-out nodes do not impact the overall signal. Note that the dropout process will randomly turn-off different nodes for each mini-batch. Dropout is more effective in reducing overfitting in deep learning than L1 or L2 regularizations.

12.1.7 Image Recognition Using FFNN

In this section, we will present a toy example of an image classification problem using the **keras** package. We use R in this section

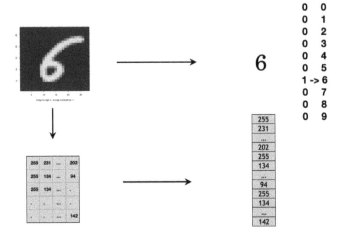

FIGURE 12.9
Grayscale image is a set of pixels on 2-d space. Each pixel has a
value range from 0 to 255

to illustrate the process and provide the python notebook on the
book website. Please check the keras R package website[1] for the
most recent development.

What is an image as data? You can consider a digital image as
a set of points in 2-d or 3-d space. For a greyscale image, each point
has a value between 0 to 255 which is considered a pixel. Figure
12.9 shows an example of a grayscale image. It is a set of pixels in
2-d space, and each pixel has a value between 0 to 255. You can
process the image as a 2-d array input using a Convolutional Neural
Network (CNN). Or, as shown in the figure, you can vectorize the
2D array into a 1D vector as the input for FFNN.

A color image is a set of pixels in 3-d space, and each pixel has
a value between 0 to 255 for a specific color format. There are three
2-d panels that represent the color red, blue, and green accordingly.
Similarly, You can process the image as a 3-d array. Or you can
vectorize the array as shown in figure 12.10.

---

[1]https://keras.rstudio.com/

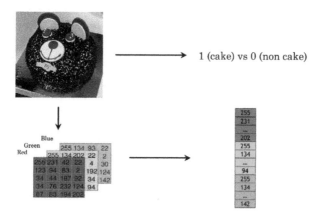

FIGURE 12.10
Color image is a set of pixels on 3-d space. Each pixel has a value range from 0 to 255

Let's look at how to use the `keras` R package for a toy example in deep learning with the handwritten digits image dataset (i.e., MNIST). `keras` has many dependent packages, so it takes a few minutes to install.

```
install.packages("keras")
```

As `keras` is just an interface to popular deep learning frameworks, we have to install the deep learning backend. The default and recommended backend is TensorFlow. By calling `install_keras()`, it will install all the needed dependencies for TensorFlow.

```
library(keras)
install_keras()
```

You can run the code in this section in the Databrick community edition with R as the interface. Refer to section 4.3 for how to set up an account, create a notebook (R or Python), and start a cluster. For an audience with a statistical background, using a well-managed cloud environment has the following benefit:

- Minimum language barrier in coding for most statisticians
- Zero setups to save time using the cloud environment
- Get familiar with the current trend of cloud computing in the industrial context

You can also run the code on your local machine with R and the required Python packages (`keras` uses the Python TensorFlow backend engine). Different versions of Python may cause some errors when running `install_keras()`. Here are the things you could do when you encounter the Python backend issue in your local machine:

- Run `reticulate::py_config()` to check the current Python configuration to see if anything needs to be changed.
- By default, `install_keras()` uses virtual environment `~/.virtualenvs/r-reticulate`. If you don't know how to set the right environment, try to set the installation method as conda (`install_keras(method = "conda")`)
- Refer to this document for more details on how to install `keras` and the TensorFlow backend[2].

Now we are all set to explore deep learning! As simple as three lines of R code, but there are quite a lot going on behind the scene. If you are using a cloud environment, you do not need to worry about these behind scene setup and maintenance.

We will use the widely used MNIST handwritten digit image dataset. More information about the dataset and benchmark results from various machine learning methods can be found at `http://yann.lecun.com/exdb/mnist/` and `https://en.wikipedia.org/wiki/MNIST_database`.

This dataset is already included in the keras/TensorFlow installation, and we can load the dataset as described in the following cell. It takes less than a minute to load the dataset.

---

[2]`https://tensorflow.rstudio.com/reference/keras/install_keras/`

```
mnist <- dataset_mnist()
```

The data structure of the MNIST dataset is straightforward and well prepared for R, which has two pieces:

(1) training set: x (i.e., features): 60000 × 28 × 28 tensor, which corresponds to 60000 28 × 28 pixel greyscale images (i.e., all the values are integers between 0 and 255 in each 28 × 28 matrix), and y (i.e., responses): a length 60000 vector which contains the corresponding digits with integer values between 0 and 9.

(2) testing set: same as the training set, but with only 10000 images and responses. The detailed structure for the dataset can be seen with str(mnist) below.

```
str(mnist)
```

```
List of 2
 $ train:List of 2
  ..$ x: int [1:60000, 1:28, 1:28] 0 0 0 0 0 0 0 0 0 0 ...
  ..$ y: int [1:60000(1d)] 5 0 4 1 9 2 1 3 1 4 ...
 $ test :List of 2
  ..$ x: int [1:10000, 1:28, 1:28] 0 0 0 0 0 0 0 0 0 0 ...
  ..$ y: int [1:10000(1d)] 7 2 1 0 4 1 4 9 5 9 ...
```

Now we prepare the features (x) and the response variable (y) for both the training and testing dataset, and we can check the structure of the x_train and y_train using str() function.

```
x_train <- mnist$train$x
y_train <- mnist$train$y
x_test <- mnist$test$x
y_test <- mnist$test$y
```

```
str(x_train)
str(y_train)
```

```
int [1:60000, 1:28, 1:28] 0 0 0 0 0 0 0 0 0 0 ...
int [1:60000(1d)] 5 0 4 1 9 2 1 3 1 4 ...
```

Now let's plot a chosen 28×28 matrix as an image using R's image() function. In R's image() function, the way of showing an image is rotated 90 degree from the matrix representation. So there are additional steps to rearrange the matrix to use image() function to show it in the actual orientation.

```
index_image = 28   ## change this index to see different image.
input_matrix <- x_train[index_image, 1:28, 1:28]
output_matrix <- apply(input_matrix, 2, rev)
output_matrix <- t(output_matrix)
image(1:28, 1:28, output_matrix, col = gray.colors(256),
    xlab = paste("Image for digit of: ", y_train[index_image]),
    ylab = "")
```

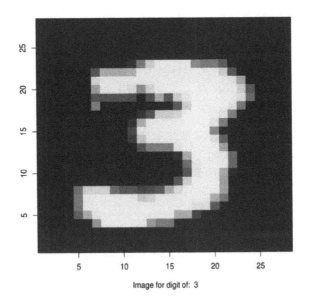

Image for digit of: 3

Here is the original 28×28 matrix for the above image:

```
dplyr::tibble(input_matrix)
```

```
## # A tibble: 28 × 1
##    input_matrix[,1]  [,2]  [,3]  [,4]  [,5]  [,6]  [,7]  [,8]  [,9] [,10] [,11]
##               <int> <int> <int> <int> <int> <int> <int> <int> <int> <int> <int>
## 1                 0     0     0     0     0     0     0     0     0     0     0
## 2                 0     0     0     0     0     0     0     0     0     0     0
## 3                 0     0     0     0     0     0     0     0     0     0     0
## 4                 0     0     0     0     0     0     0     0     0     0     0
## 5                 0     0     0     0     0     0     0     0     0     0     0
## 6                 0     0     0     0     0     0     0     0     0     0     9
## # … with 22 more rows, and 1 more variable: input_matrix[12:28] <int>
...
```

There are multiple deep learning methods for solving the hand-written digits problem and we will start with the simple and generic one, feedforward neural network (FFNN). FFNN contains a few fully connected layers and information flows from the front layer to a back layer without any feedback loop from the back layer to the front layer. It is the most common deep learning model to start with.

12.1.7.1 Data Preprocessing

In this section, we will walk through the steps required for data preprocessing. For the MNIST dataset that we just loaded has already undergone some preprocessing, so it is relatively "clean". However, before we feed the data into FFNN, we still need to do some additional preparations.

First, for each digit, we have a scalar response and a 28×28 integer matrix with values between 0 and 255. To use the out-of-box DNN functions, each response requires all its features to be in a single row (i.e., a vector). For example, an image in the MNIST dataet has an input of a 28×28 matrix for one response (y), not a single row of many columns. We need to convet the 28×28 matrix into a single row by appending every row of the matrix to the first row using reshape() function.

In addition, we also need to scale all features to be between $(0, 1)$ or $(-1, 1)$ or close to $(-1, 1)$ range. This will improve numerical stability in the optimization procedure, as there are a lot of parameters to be optimized.

We first reshape the 28×28 image for each digit (i.e each row) into 784 columns (i.e. features), and then rescale the value to be between 0 and 1 by dividing the original pixel value by 255, as described in the cell below.

```
# step 1: reshape
x_train <- array_reshape(x_train,
                         c(nrow(x_train), 784))
x_test <- array_reshape(x_test,
                        c(nrow(x_test), 784))

# step 2: rescale
x_train <- x_train / 255
x_test <- x_test / 255
```

And here is the structure of the reshaped and rescaled features for training and testing dataset. Now for each digit, there are 784 columns of features.

```
str(x_train)
str(x_test)
```

```
num [1:60000, 1:784] 0 0 0 0 0 0 0 0 0 0 ...
num [1:10000, 1:784] 0 0 0 0 0 0 0 0 0 0 ...
```

In this example, though the response variable is an integer (i.e. the corresponding digits for an image), there is no order or rank for these integers and they are just an indication of one of the 10 categories. We convert the response variable y to be categorical.

```
y_train <- to_categorical(y_train, 10)
y_test <- to_categorical(y_test, 10)
str(y_train)
```

```
num [1:60000, 1:10] 0 1 0 0 0 0 0 0 0 0 ...
```

12.1.7.2 Fit Model

Now we are ready to fit the model. It is straightforward to build a deep neural network using keras. For this example, the number of input features is 784 (i.e. the scaled value of each pixel in the 28×28 image) and the number of class for the output is 10 (i.e. one of the ten categories). So the input size for the first layer is 784 and the output size for the last layer is 10. And we can add any number of compatible layers in between.

In keras, it is easy to define a DNN model: (1) use `keras_model_sequential()` to initiate a model placeholder and all model structures are attached to this model object, (2) layers are added in sequence by calling the `layer_dense()` function, (3) add arbitrary layers to the model based on the sequence of calling `layer_dense()`.

For a dense layer, all nodes from the previous layer are connected with each and every node to the next layer. In the `layer_dense()` function, we can define how many nodes in that layer through the `units` parameter. The activation function can be defined through the `activation` parameter. For the first layer, we also need to define the input features' dimension through `input_shape` parameter. For our preprocessed MNIST dataset, there are 784 columns in the input data. To reduce overfitting, use the dropout method, which randomly drops a proportion of the nodes in a layer. Define the dropout proportion through `layer_dropout()` function immediately after the `layer_dense()` function.

```
dnn_model <- keras_model_sequential()
dnn_model %>%
  layer_dense(units = 256, activation = 'relu', input_shape = c(784)) %>%
  layer_dropout(rate = 0.4) %>%
  layer_dense(units = 128, activation = 'relu') %>%
  layer_dropout(rate = 0.3) %>%
  layer_dense(units = 64, activation = 'relu') %>%
  layer_dropout(rate = 0.3) %>%
  layer_dense(units = 10, activation = 'softmax')
```

The above dnn_model has 4 layers with first layer 256 nodes, 2nd layer 128 nodes, 3rd layer 64 nodes, and last layer 10 nodes. The activation function for the first 3 layers is relu and the activation function for the last layer is softmax which is typical for classification problems. The model detail can be obtained through summary() function. The number of parameter of each layer can be calculated as: (number of input features +1) times (numbe of nodes in the layer). For example, the first layer has $(784 + 1) \times 256 = 200960$ parameters; the 2nd layer has $(256 + 1) \times 128 = 32896$ parameters. Please note, dropout only randomly drop a certain proportion of parameters for each batch; it does not reduce the number of parameters in the model. The total number of parameters for the dnn_model we just defined is 242762.

```
summary(dnn_model)
```

Layer (type)	Output Shape	Param #
dense_1 (Dense)	(None, 256)	200960
dropout_1 (Dropout)	(None, 256)	0
dense_2 (Dense)	(None, 128)	32896

```
dropout_2 (Dropout)              (None, 128)                    0
-----------------------------------------------------------------------
dense_3 (Dense)                  (None, 64)                  8256
-----------------------------------------------------------------------
dropout_3 (Dropout)              (None, 64)                     0
-----------------------------------------------------------------------
dense_4 (Dense)                  (None, 10)                   650
=======================================================================
Total params: 242,762
Trainable params: 242,762
Non-trainable params: 0
-----------------------------------------------------------------------
```

Once a model is defined, we need to compile the model with a few other hyper-parameters, including:

1. Loss function

2. Optimizer

3. Performance metrics

For multi-class classification problems, people usually use `categorical_crossentropy` the loss function and `optimizer_rmsprop()` as the optimizer which performs batch gradient descent.

```
dnn_model %>% compile(
  loss = 'categorical_crossentropy',
  optimizer = optimizer_rmsprop(),
  metrics = c('accuracy')
)
```

Now we can feed data (x and y) into the neural network we just built to estimate all the parameters in the model. Here we define three hyperparameters: `epochs`, `batch_size`, and `validation_split`, for this model. It only takes a couple of minutes to complete.

```
dnn_history <- dnn_model %>% fit(
  x_train, y_train,
  epochs = 15, batch_size = 128,
  validation_split = 0.2
)
```

There is some useful information stored in the output object dnn_history and the details can be shown by using str(). We can plot the training and validation accuracy and loss as function of epoch by calling plot(dnn_history).

```
str(dnn_history)
```

```
List of 2
 $ params :List of 8
  ..$ metrics            : chr [1:4] "loss" "acc" "val_loss" "val_acc"
  ..$ epochs             : int 15
  ..$ steps              : NULL
  ..$ do_validation      : logi TRUE
  ..$ samples            : int 48000
  ..$ batch_size         : int 128
  ..$ verbose            : int 1
  ..$ validation_samples : int 12000
 $ metrics:List of 4
  ..$ acc     : num [1:15] 0.83 0.929 0.945 0.954 0.959 ...
  ..$ loss    : num [1:15] 0.559 0.254 0.195 0.165 0.148 ...
  ..$ val_acc : num [1:15] 0.946 0.961 0.967 0.969 0.973 ...
  ..$ val_loss: num [1:15] 0.182 0.137 0.122 0.113 0.104 ...
 - attr(*, "class")= chr "keras_training_history"
```

```
plot(dnn_history)
```

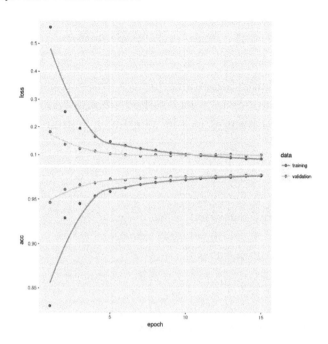

12.1.7.3 Prediction

```
dnn_model %>% evaluate(x_test, y_test)
```

```
##         loss    accuracy
## 0.09035096 0.98100007
```

```
dnn_pred <- dnn_model %>%
  predict(x_test) %>%
  k_argmax()
head(dnn_pred)
```

```
tf.Tensor([7 2 1 0 4 1], shape=(6,), dtype=int64)
```

Let's check a few misclassified images. We can use the following code to find a number of misclassified images. And we can plot these misclassified images to see whether a human can correctly read it out.

```
## Convert tf.tensor to array
dnn_pred <- as.array(dnn_pred)
## total number of mis-classcified images
sum(dnn_pred != mnist$test$y)
```

```
[1] 190
```

```
missed_image = mnist$test$x[dnn_pred != mnist$test$y,,]
missed_digit = mnist$test$y[dnn_pred != mnist$test$y]
missed_pred = dnn_pred[dnn_pred != mnist$test$y]

index_image = 34

## change this index to see different image.
input_matrix <- missed_image[index_image,1:28,1:28]
output_matrix <- apply(input_matrix, 2, rev)
output_matrix <- t(output_matrix)

image(1:28, 1:28, output_matrix, col = gray.colors(256),
    xlab = paste("Image for digit ", missed_digit[index_image],
        ", wrongly predicted as ", missed_pred[index_image]),
    ylab = "")
```

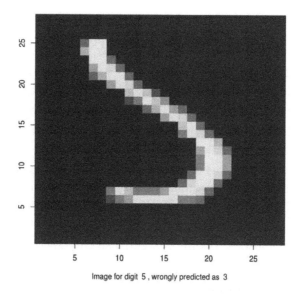

Image for digit 5 , wrongly predicted as 3

Now we finish this simple tutorial of using deep neural networks for handwritten digit recognition using the MNIST dataset. We illustrate how to:

- Reshape the original data into the right format and scale it

- Define a deep neural network with an arbitrary number of layers

- Choose an activation function, optimizer, and loss function

- Use dropout to limit overfitting

- Setup hyperparameters

- Fit the model and use a fitted model to predict

Finally, we illustrate how to plot the accuracy/loss as functions of the epoch. This shows the end-to-end cycle of how to fit a deep neural network model.

On the other hand, the image can be better dealt with Convolutional Neural Network (CNN) and we are going to walk through the exact same problem using CNN in the next section.

12.2   Convolutional Neural Network

Using a feedforward neural network to solve computer vision problems presents several challenges. One of the challenges is that the input vector can become very large after you vectorize the image array. A $64 \times 64 \times 3$ image results in an input vector of 12288! As the image size increases, the number of parameters grows quickly and it can be difficult to get enough data to fit the model. Additionally, the computational requirements to train a feedforward neural network will soon become infeasible. Furthermore, vectorization causes the loss of most of the spacial information of the image. To overcome these issues, people instead use the convolutional neural network for computer vision problems.

This section introduces the Convolutional Neural Network (CNN), a deep learning model that is almost universally used in computer vision applications. Computer vision has been advancing rapidly which enables many new applications such as self-driving cars and unlocking a phone using face recognition. The application is not limited to the tech industry, but also extends to traditional industries such as agriculture and healthcare.

For example, precision agriculture utilizes advanced hardware and computer vision algorithms to increase efficiency and reduce costs for farmers. Images from cameras in the greenhouse can be analyzed to track plant growth state, while sensory data from drones, satellites, and tractors can be used to track soil conditions, detect herbs and pests, automate irrigation, etc. In health care, computer vision helps clinicians to diagnose disease, and identify cancer sites with high accuracy (Kwak and Hui, 2019). Even if you don't work on computer vision, you may find some of the ideas inspiring and borrow them into your area.

Some popular computer vision problems are

- Image classification (or image recognition): Recognize the object in the image. Is there a cat in the image? Who is this person?

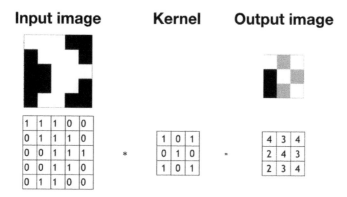

FIGURE 12.11
There are an input image (left), a filter (middel), and an output image (right)

- Object detection: Detect the position and boarder of a specific object. For example, if you are building a self-driving car, you need to know the positions of other cars around you.
- Neural Style Transfer (NST): Given a "content" image and a "style" image, generate an image that merges the two.

12.2.1 Convolution Layer

A fundamental building block of the convolution neural network is, as the name indicates, the convolution operation. In this chapter, we illustrate the fundamentals of CNN using the example of image classification.

How do you do convolution? For example, you have a 5×5 2-d image (figure 12.11). You apply a 3×3 filter and convolute over the image. The output of this convolution operator will be a 3×3 matrix, which you can consider as a 3×3 image and visualize it (top right of figure 12.11).

Starting from the top left corner of the image, place the filter on the top left 3×3 sub-matrix of the input image and take the element-wise product. Then, add up the 9 numbers. Move one step forward each time until it gets to the bottom right. The detailed process is shown in figure 12.12.

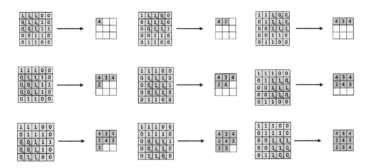

FIGURE 12.12

Convolution step by step

FIGURE 12.13

Edge detection example

Let's use edge detection as an example to illustrate how convolution operation works. Given a picture as the left of figure 12.13, you want to detect the vertical lines. For example, there are vertical lines along with the hair and the edges of the bookcase. How do you do that? There are standard filters for operations like blurring, sharpening, and edge detection. To get the edge, you can use the following 3×3 filter to convolute over the image.

```
kernel_vertical = matrix(c(1, 1, 1, 0, 0, 0, -1, -1, -1),
nrow = 3, ncol = 3)

kernel_vertical
```

```
##        [,1] [,2] [,3]
## [1,]    1    0   -1
## [2,]    1    0   -1
## [3,]    1    0   -1
```

The following code implements the convolution process. The result is shown as the middle of figure 12.13.

```
image = magick::image_read("http://bit.ly/2Nh5ANX")
kernel_vertical = matrix(c(1, 1, 1, 0, 0, 0, -1, -1, -1),
                         nrow = 3, ncol = 3)

kernel_horizontal = matrix(c(1, 1, 1, 0, 0, 0, -1, -1, -1),
                           nrow = 3, ncol = 3, byrow = T)

image_edge_vertical = magick::image_convolve(image, kernel_vertical)
image_edge_horizontal = magick::image_convolve(image, kernel_horizontal)

par(mfrow = c(1, 3))

plot(image)
plot(image_edge_vertical)
plot(image_edge_horizontal)
```

Why can kernel_vertical detect vertical edge? Let's look at a simpler example. The following 8 × 8 matrix where half of the matrix is 10 and the other half is 0. The corresponding image is shown as the left of the figure 12.14.

```
input_image = matrix(rep(c(200, 200, 200, 200, 0, 0, 0, 0), 8),
                      nrow = 8, byrow = T)
input_image
```

```
##        [,1] [,2] [,3] [,4] [,5] [,6] [,7] [,8]
## [1,]   200  200  200  200    0    0    0    0
## [2,]   200  200  200  200    0    0    0    0
```

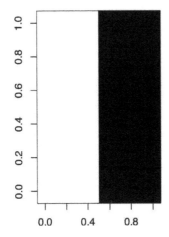 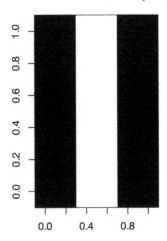

FIGURE 12.14
Simple edge detection example

```
## [3,]  200  200  200  200   0   0   0   0
## [4,]  200  200  200  200   0   0   0   0
## [5,]  200  200  200  200   0   0   0   0
## [6,]  200  200  200  200   0   0   0   0
## [7,]  200  200  200  200   0   0   0   0
## [8,]  200  200  200  200   0   0   0   0
```

If we use the above filter kernel_vertical, the output matrix is shown below and the corresponding image is shown as the right of the figure 12.14.

```
output_image = matrix(rep(c(0, 0, 200, 200, 0, 0), 6),
                      nrow = 6, byrow = T)
output_image
```

```
##         [,1] [,2] [,3] [,4] [,5] [,6]
## [1,]      0    0  200  200    0    0
## [2,]      0    0  200  200    0    0
## [3,]      0    0  200  200    0    0
## [4,]      0    0  200  200    0    0
## [5,]      0    0  200  200    0    0
## [6,]      0    0  200  200    0    0
```

So the output image has a lighter region in the middle that corresponds to the vertical edge of the input image. When the input image is large, such as the image in figure 12.13 is 1020×711, the edge will not seem as thick as it is in this small example. To detect the horizontal edge, you only need to rotate the filter by 90 degrees. The right image in figure 12.13 shows the horizontal edge detection result. You can see how convolution operator detects a specific feature from the image.

The parameters for the convolution operation are the elements in the filter. For a 3×3 filter shown below, the parameters to estimate are w_1 to w_9. So far, we move the filter one step each time when we convolve. You can do more than 1 step as well. For example, you can hop 2 steps each time after the sum of the element-wise product. It is called strided-convolution. Use stride s means the output is downsized by a factor of s. It is rarely used in practice but it is good to be familiar with the concept.

12.2.2 Padding layer

Assume the stride is s and we have a $n \times n$ input image to convolve with a $f \times f$ filter, the output image is $(\frac{n-f}{s}+1) \times (\frac{n-f}{s}+1)$. After each convolution, the dimension of the output shrinks. Depending on the size of the input image, the output size may get too small after a few rounds. Also, the pixel at the corner is used less than the pixel in the middle. So it overlooks some information in the image. To overcome these problems, you can add a padding layer during the process. To keep the output the same size as the input image in figure 12.15, you can pad two pixels on each side of the image with 0 (figure 12.16).

If the stride is s and we have a $n \times n$ input image to convolve with a $f \times f$ filter. This time we pad p pixels in each side, then the output size becomes $(\frac{n+2p-f}{s}+1) \times (\frac{n+2p-f}{s}+1)$. You can specify the value for p and also the pixel value used. Or you can just use 0 to pad and make the output the same size with input.

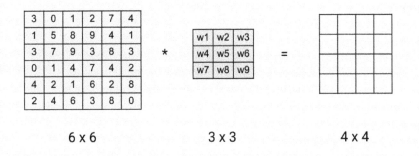

FIGURE 12.15
Convolution layer parameters

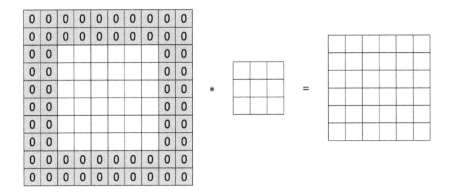

FIGURE 12.16
Padding layer

12.2.3 Pooling Layer

People sometimes use the pooling layer to reduce the size of the representation and make some of the feature detection more robust. If you have a 4×4 input, the max and mean pooling operation are shown in the figure 12.17. The process is quite simple. In the example, the filter is 2×2 and stride is 2, so break the input into four 2×2 regions (shown in the figure with different shaded colors). For max pooling, each of the outputs is the maximum from the

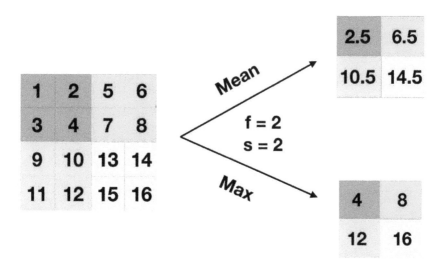

FIGURE 12.17
Pooling layer

corresponding shaded sub-region. Mean pooling layer works in the same way except for getting the mean instead of maximum of the sub-region. The pooling layer has hyperparameters (f and s) but it has no parameter for gradient descent to learn.

Let's go through an example of pooling a 2D grayscale image. Hope it gives you some intuition behind what it does. Read the image and convert the original color image (a 3D array) to grayscale (a 2D matrix).

```r
library(EBImage)
library(dplyr)

eggshell <- readImage("https://scientistcafe.com/images/eggshell.jpeg") %>%
  # make it smaller
  resize(560, 420) %>%
  # rotate image
  rotate(90)
```

```
# convert to 2D grayscale
gray_eggshell = apply(eggshell, c(1,2), mean)
```

The following function takes an image matrix or array, and apply pooling operation.

```
pooling <- function(type = "max", image, filter, stride) {
    f <- filter
    s <- stride

    if (length(dim(image)) == 3) {
        # get image dimensions
        col <- dim(image[, , 1])[2]
        row <- dim(image[, , 1])[1]
        # calculate new dimension size
        c <- (col - f)/s + 1
        r <- (row - f)/s + 1
        # create new image object
        newImage <- array(0, c(c, r, 3))
        # loops in RGB layers
        for (rgb in 1:3) {
            m <- image[, , rgb]
            m3 <- matrix(0, ncol = c, nrow = r)
            i <- 1
            if (type == "mean")
                for (ii in 1:r) {
                    j <- 1
                    for (jj in 1:c) {
                        m3[ii, jj] <- mean(as.numeric(m[i:(i +
                            (f - 1)), j:(j + (f - 1))]))
                        j <- j + s
                    }
                    i <- i + s
                } else for (ii in 1:r) {
                    j = 1
```

```
            for (jj in 1:c) {
              m3[ii, jj] <- max(as.numeric(m[i:(i +
                (f - 1)), j:(j + (f - 1))]))
              j <- j + s
            }
            i <- i + s
          }
        newImage[, , rgb] <- m3
      }
  } else if (length(dim(image)) == 2) {
    # get image dimensions
    col <- dim(image)[2]
    row <- dim(image)[1]
    # calculate new dimension size
    c <- (col - f)/s + 1
    r <- (row - f)/s + 1
    m3 <- matrix(0, ncol = c, nrow = r)
    i <- 1
    if (type == "mean")
        for (ii in 1:r) {
            j <- 1
            for (jj in 1:c) {
              m3[ii, jj] <- mean(as.numeric(image[i:(i +
                (f - 1)), j:(j + (f - 1))]))
              j <- j + s
            }
            i <- i + s
        } else for (ii in 1:r) {
        j = 1
        for (jj in 1:c) {
            m3[ii, jj] <- max(as.numeric(image[i:(i +
              (f - 1)), j:(j + (f - 1))]))
            j <- j + s
        }
        i <- i + s
    }
```

```
        newImage <- m3
    }
    return(newImage)
}
```

Let's apply both max and mean pooling with filter size 10 ($f = 10$) and stride 10 ($s = 10$).

```
gray_eggshell_max = pooling(type = "max",
                            image = gray_eggshell,
                            filter = 10, stride = 10)

gray_eggshell_mean = pooling(type = "mean",
                             image = gray_eggshell,
                             filter = 10, stride = 10)
```

You can see the result by plotting the output image (figure 12.18). The top left is the original color picture. The top right is the 2D grayscale picture. The bottom left is the result of max pooling. The bottom right is the result of mean pooling. The max-pooling gives you the value of the largest pixel and the mean-pooling gives the average of the patch. You can consider it as a representation of features, looking at the maximal or average presence of different features. In general, max-pooling works better. You can gain some intuition from the example (figure 12.18). The max-pooling "picks" more distinct features and average-pooling blurs out features evenly.

```
par(mfrow = c(2,2), oma = c(1, 1, 1, 1))
plot(eggshell)
plot(as.Image(gray_eggshell))
plot(as.Image(gray_eggshell_max))
plot(as.Image(gray_eggshell_mean))
```

FIGURE 12.18

Example of max and mean pooling

12.2.4 Convolution Over Volume

So far, we have shown different types of layers on 2D inputs. If you have a 3D input (such as a color image), then the filters will have 3 channels too. For example, if you have a 6×6 color image, the input dimension is $6 \times 6 \times 3$. We call them the height, width, and the number of channels. The filter itself has 3 channels corresponding to the red, green, and blue channels of the input. You can consider it as a 3D cube with 27 parameters. Apply each channel of the filter to the corresponding channel of the input. Multiply each of the 27 numbers with the corresponding numbers from the top left region of the color input image and add them up. Add a bias parameter and apply an activation function which gives you the first number of the output image. Then slide it over to calculate the next one. The final output is 2D 4×4. If you want to detect features in the red channel only, you can use a filter with the second and third channels to be all 0s. With different choices of the parameters, you

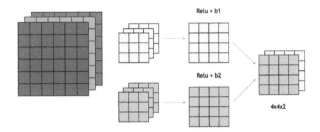

FIGURE 12.19

Example of convolution over volume

can get different feature detectors. You can use more than one filter and each filter has multiple channels. For example, you can use one $3 \times 3 \times 3$ filter to detect the horizontal edge and another to detect the vertical edge. Figure 12.18 shows an example of one layer with two filters. Each filter has a dimension of $3 \times 3 \times 3$. The output dimension is $4 \times 4 \times 2$. The output has two channels because we have two filters on the layer. The total number of parameters is 58 (each filter has one bias parameter b).

We use the following notations for layer l:

- Use $n_W^{[l]}$, $n_H^{[l]}$, and $n_C^{[l]}$ to denote the width, height and number of channels of the input
- $f^{[l]}$ is the filter size
- $p^{[l]}$ is the padding size
- $s^{[l]}$ is the stride

The number of filters for layer l is the number of channels of the output of layer l. Since the output of layer l is the input of layer $l + 1$, the number of filters for layer l is $n_C^{[l+1]}$. The number of channels of the filter of layer l should be equal to the number of channels of the input of layer l which is the output of layer $l - 1$ (i.e. $n_C^{[l-1]}$). So the dimensions of some key elements of layer l are

- Each filter: $f^{[l]} \times f^{[l]} \times n_C^{[l-1]}$
- Activations: $a^{[l]} \to n_H^{[l]} \times n_W^{[l]} \times n_C^{[l]}$

- Weights: $f^{[l]} \times f^{[l]} \times n_C^{[l-1]} \times n_C^{[l]}$
- bias: $b_C^{[l]}$
- Input: $n_H^{[l-1]} \times n_W^{[l-1]} \times n_C^{[l-1]}$
- Output: $n_H^{[l]} \times n_W^{[l]} \times n_C^{[l]}$

After a series of 3D convolutional layers, we need to "flatten" the 3D tensor to a 1D tensor, and add one or several dense layers to connect the output to the response variable.

Now you know the basic building blocks of CNN. Let's look at how to use the keras R package to solve the same handwritten digits image recognition problem as in section 12.1.7. You will see the CNN is better at handling image recognition problem.

12.2.5 Image Recognition Using CNN

CNN leverages the relationship among neighbor pixels in the 2D image for better performance. It also avoids generating thousands or millions of features for high resolution images with full color. Now let's import the MNIST dataset again as we have done some preprocessing specifically for FFNN before. CNN requires different preprocessing steps. Let's start with a few parameters to be used later.

```
# Load the mnist data's training and testing dataset
mnist <- dataset_mnist()
x_train <- mnist$train$x
y_train <- mnist$train$y
x_test <- mnist$test$x
y_test <- mnist$test$y


# Define a few parameters to be used in the CNN model
batch_size <- 128
num_classes <- 10
epochs <- 10


# Input image dimensions
```

```
img_rows <- 28
img_cols <- 28
```

12.2.5.1 Data preprocessing

For CNN, the input is a $n_H \times n_W \times n_C$ 3D array with n_C channels. For example, a greyscale $n_H \times n_W$ image has only one channel, and the input is $n_H \times n_W \times 1$ tensor. A $n_H \times n_W$ 8-bit per channel RGB image has three channels with 3 $n_H \times n_W$ array with values between 0 and 255, so the input is $n_H \times n_W \times 3$ tensor. For the problem that we have now, the image is greyscaled, but we need to specifically define there is one channel by reshaping the 2D array into 3D tensor using `array_reshape()`. The `input_shape` variable will be used in the CNN model later. For an RGB color image, the number of channels is 3 and we need to replace "1" with "3" for the code cell below if the input image is RGB format.

```
x_train <- array_reshape(x_train,
                         c(nrow(x_train), img_rows, img_cols, 1))
x_test <- array_reshape(x_test,
                        c(nrow(x_test), img_rows, img_cols, 1))
input_shape <- c(img_rows, img_cols, 1)
```

Here is the structure of the reshaped image, the first dimension is the image index, the 2-4 dimension is a 3D tensor even though there is only one channel.

```
str(x_train)
```

```
int [1:60000, 1:28, 1:28, 1] 0 0 0 0 0 0 0 0 0 0 ...
```

Same as the FFNN model, we scale the input values to be between 0 and 1 for the same numerical stability consideration in the optimization process.

```
x_train <- x_train / 255
x_test <- x_test / 255
```

Encode the response variable to binary vectors.

```
# Convert class vectors to binary class matrices
y_train <- to_categorical(y_train, num_classes)
y_test <- to_categorical(y_test, num_classes)
```

12.2.5.2 Fit Model

CNN model contains a series of 3D convolutional layers which contains a few parameters:

(1) the kernal_size which is typically 3×3 or 5×5;

(2) the number of filters, which is equal to the number of channels of the output;

(3) activation function.

For the first layer, there is also an input_shape parameter which is the input image size and channel. To prevent overfitting and speed up computation, a pooling layer is usually applied after one or a few convolutional layers. A maximum pooling layer with pool_size $= 2 \times 2$ reduces the size to half. Dropout can be used as well in addition to pooling neighbor values. After a few 3D convolutional layers, we also need to "flatten" the 3D tensor output into 1D tensor, and then add one or a couple of dense layers to connect the output to the target response classes.

Let's define the CNN model structure. Now we define a CNN model with two convolutional layers, two max-pooling layers, and two dropout layers to mediate overfitting. There are three dense layers after flattening the 3D tensor. The last layer is a dense layer that connects to the response.

```
# define model structure
cnn_model <- keras_model_sequential() %>%
  layer_conv_2d(filters = 32,
                kernel_size = c(5,5),
                activation = 'relu',
                input_shape = input_shape) %>%
  layer_max_pooling_2d(pool_size = c(2, 2)) %>%
  layer_conv_2d(filters = 64,
                kernel_size = c(5,5),
                activation = 'relu') %>%
  layer_max_pooling_2d(pool_size = c(2, 2)) %>%
  layer_dropout(rate = 0.2) %>%
  layer_flatten() %>%
  layer_dense(units = 120, activation = 'relu') %>%
  layer_dropout(rate = 0.5) %>%
  layer_dense(units = 84, activation = 'relu') %>%
  layer_dense(units = num_classes, activation = 'softmax')


summary(cnn_model)
```

```
## Model: "sequential"
##
## _____
## Layer (type)                        Output Shape               Param #
## ========================================================================
## conv2d (Conv2D)                     (None, 24, 24, 32)          832
##
## max_pooling2d (MaxPooling2D)        (None, 12, 12, 32)          0
##
## conv2d_1 (Conv2D)                   (None, 8, 8, 64)            51264
##
## max_pooling2d_1 (MaxPooling2D)      (None, 4, 4, 64)            0
##
## dropout (Dropout)                   (None, 4, 4, 64)            0
##
## flatten (Flatten)                   (None, 1024)                0
##
## dense (Dense)                       (None, 120)                 123000
##
## dropout_1 (Dropout)                 (None, 120)                 0
##
## dense_1 (Dense)                     (None, 84)                  10164
##
## dense_2 (Dense)                     (None, 10)                  850
## ========================================================================
## Total params: 186,110
## Trainable params: 186,110
## Non-trainable params: 0
## _____
```

Similar to before, we need to compile the defined CNN model.

```
cnn_model %>% compile(
  loss = loss_categorical_crossentropy,
  optimizer = optimizer_adadelta(),
  metrics = c('accuracy')
)
```

We then train the model and save each epochs's history using fit() function. Please note, as we are not using GPU, it takes a few minutes to finish. Please be patient while waiting for the results. The training time can be significantly reduced if running on GPU.

```
cnn_history <- cnn_model %>% fit(
  x_train, y_train,
  batch_size = batch_size,
  epochs = epochs,
```

```
validation_split = 0.2
)
```

The trained model accuracy can be evaluated on the testing dataset which is pretty good.

```
cnn_model %>% evaluate(x_test, y_test)
```

```
##       loss   accuracy
## 0.02301287 0.99300003
```

There is some useful information stored in the output object `cnn_history` and the details can be shown by using `str()`. We can plot the training and validation accuracy and loss as function of epoch by simply calling `plot(cnn_history)`.

```
str(cnn_history)
```

```
## List of 2
##  $ params :List of 3
##   ..$ verbose: int 1
##   ..$ epochs : int 10
##   ..$ steps  : int 375
##  $ metrics:List of 4
##   ..$ loss        : num [1:10] 0.3415 0.0911 0.0648 0.0504 0.0428 ...
##   ..$ accuracy    : num [1:10] 0.891 0.973 0.981 0.985 0.987 ...
##   ..$ val_loss    : num [1:10] 0.071 0.0515 0.0417 0.0377 0.0412 ...
##   ..$ val_accuracy: num [1:10] 0.978 0.985 0.988 0.99 0.988 ...
##  - attr(*, "class")= chr "keras_training_history"
```

```
plot(cnn_history)
```

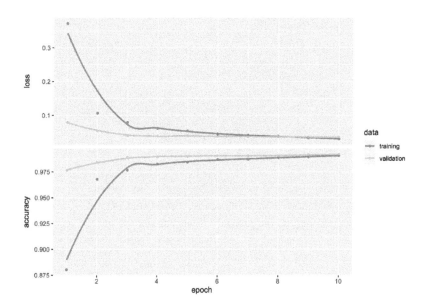

12.2.5.3 Prediction

We can apply the trained model to predict new image.

```
# model prediction
cnn_pred <- cnn_model %>%
  predict(x_test) %>%
  k_argmax()

head(cnn_pred)
```

```
## tf.Tensor([7 2 1 0 4 1], shape=(6,), dtype=int64)
```

Now let's check a few misclassified images to see whether a human can do a better job than this simple CNN model.

```
## Convert tf.tensor to array
cnn_pred <- as.array(cnn_pred)
## number of mis-classcified images
sum(cnn_pred != mnist$test$y)
```

```
## [1] 70
```

```
missed_image = mnist$test$x[cnn_pred != mnist$test$y,,]
missed_digit = mnist$test$y[cnn_pred != mnist$test$y]
missed_pred = cnn_pred[cnn_pred != mnist$test$y]
```

```
index_image = 10 ## change this index to see different image.
input_matrix <- missed_image[index_image,1:28,1:28]
output_matrix <- apply(input_matrix, 2, rev)
output_matrix <- t(output_matrix)
image(1:28, 1:28, output_matrix, col=gray.colors(256),
xlab=paste('Image for digit ', missed_digit[index_image], ',
wrongly predicted as ', missed_pred[index_image]), ylab="")
```

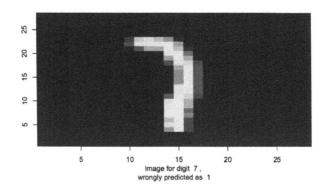

Image for digit 7 ,
wrongly predicted as 1

12.3 Recurrent Neural Network

Traditional neural networks don't have a framework to handle sequential events, whereas later events are based on the previous ones. For example, map an input audio clip to a text transcript where the input is voice over time, and the output is the corresponding sequence of words over time. Recurrent Neural Network is a deep-learning model that can process this type of sequential data.

The recurrent neural network allows information to flow from one step to the next with a repetitive structure. Figure 12.20 shows the basic chunk of an RNN network. You combine the activated neuro from the previous step with the current input $x^{<t>}$ to produce an output $\hat{y}^{<t>}$ and an updated activated neuro to support the next input at $t + 1$.

So the whole process repeats a similar pattern. If we unroll the loop (figure 12.21, the chain-like recurrent nature makes it the natural architecture for sequential data.

There is incredible success applying RNNs to this type of problems:

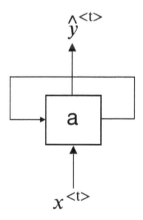

FIGURE 12.20
Recurrent neural network unit

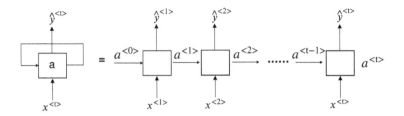

FIGURE 12.21

An unrolled recurrent neural network

- Machine translation
- Voice recognition
- Music generation
- Sentiment analysis

A trained CNN accepts a fixed-sized vector as input (such as 28×28 image) and produces a fixed-sized vector as output (such as the probabilities of being one the ten digits). RNN has a much more flexible structure. It can operate over sequences of vectors and produces sequences of outputs and they can vary in size. To understand what it means, let's look at some RNN structure examples.

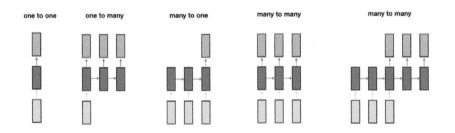

The rectangle represents a vector and the arrow represents matrix multiplications. The input vector is in green and the output vector is in blue. The red rectangle holds the intermediate state. From left to right:

- one-to-one: model takes a fixed size input and produces a fixed size output, such as CNN. it is not sequential.

- one-to-many: model takes one input and generate a sequence of output, such as the music generation.
- many-to-one: model takes a sequence of input and produces a single output, such as sentiment analysis.
- many-to-many: model takes a sequence of input and produces a sequence of output. The input size can be the same with the output size (such as name entity recognition) or it can be different (such as machine translation).

12.3.1 RNN Model

To further understand the RNN model, let's look at an entity recognition example. Assume you want to build a sequence model to recognize the company or computer language names in a sentence like this: "Use Netlify and Hugo." It is a name recognition problem which is used by the research company to index different company names in the articles. For tasks like this, we need a model that can learn and share the learning across different texts. The position of the word has important information about the word. For example, the word before "to" is more likely to be a verb than a noun. It is also used in material science to tag chemicals mentioned in the most recent journals to find any indication of the next research topic.

Given input sentence x, you want a model to produce one output for each word in x that tells you if that word is the name for something. So in this example, the input is a sequence of 5 words including the period in the end. The output is a sequence of 0/1 with the same length that indicates whether the input word is a name (1) or not (0). We use superscript $<t>$ to denote the element position of input and output; use superscript (i) to denote the ith sample (you will have different sentences in the training data); Use $T_x^{(i)}$ to denote the length of ith input, $T_y^{(i)}$ for output. In this case, $T_x^{(i)}$ is equal to $T_y^{(i)}$.

Before we build a neural network, we need to decide a way to represent individual words in numbers. What should $x^{<1>}$ be? In practice, people use word embedding which we will discuss in the later section. Here, for illustration, we use the one-hot encoding x: Use($x^{<1>}$) Netlify($x^{<2>}$) and($x^{<3>}$) Hugo($x^{<4>}$)

y: 0 ($y^{<1>}$) 1($y^{<2>}$) 0($y^{<3>}$) 1($y^{<4>}$)

$x^{(i)<t>}, T_x^{(i)}$ (i^{th} sample)

$y^{(i)<t>}, T_y^{(i)}$ (i^{th} sample)

word representation. Assume we have a dictionary of 10,000 unique words. You can build the dictionary by finding the top 10,000 occurring words in your training set. Each word in your training set will have a position in the dictionary sequence. For example, "use" is the 8320th element of the dictionary sequence. So $x^{<1>}$ is a vector with all zeros except for a one on position 8320. Using this one-hot representation, each input $x^{<t>}$ is a vector with all zeros except for one element.

$$
\begin{bmatrix} a[1] \\ aaron[2] \\ \vdots \\ and[360] \\ \vdots \\ Hugo[4075] \\ \vdots \\ Netlify[5210] \\ \vdots \\ use[8320] \\ \vdots \\ Zulu[10000] \end{bmatrix}
\Rightarrow use = \begin{bmatrix} 0 \\ 0 \\ \vdots \\ 0 \\ \vdots \\ 0 \\ \vdots \\ 0 \\ \vdots \\ 1 \\ \vdots \\ 0 \end{bmatrix},
Netlify = \begin{bmatrix} 0 \\ 0 \\ \vdots \\ 0 \\ \vdots \\ 0 \\ \vdots \\ 1 \\ \vdots \\ 0 \\ \vdots \\ 0 \end{bmatrix},
and = \begin{bmatrix} 0 \\ 0 \\ \vdots \\ 1 \\ \vdots \\ 0 \\ \vdots \\ 0 \\ \vdots \\ 0 \\ \vdots \\ 0 \end{bmatrix},
Hugo = \begin{bmatrix} 0 \\ 0 \\ \vdots \\ 0 \\ \vdots \\ 1 \\ \vdots \\ 0 \\ \vdots \\ 0 \\ \vdots \\ 0 \end{bmatrix}
$$

Given this representation of input words, the goal is to learn a sequence model that maps the input words to output y, indicating if the word is an entity (1) or not (0). Let us build a one-layer recurrent neural network. The model starts from the first word "use" ($x^{<1>}$) and build a neural network to predict the output. To start the process, we also need to initialize the activation at time 0, a_0. The most common choice is to use a vector of zeros. The common activation function for the intermediate layer is the Hyperbolic Tangent Function (tanh). RNN uses other methods to prevent the vanishing gradient problem discussed in section 12.3.2.

Similar to FFNN, The output layer activation function depends on the output type. The current example is a binary classification, so we use the sigmoid function (σ).

$$a^{<0>} = 0; \quad a^{<1>} = tanh(W_{aa}a^{<0>} + W_{ax}x^{<1>} + b_a)$$
$$\hat{y}^{<1>} = \sigma(W_{ya}a^{<1>} + b_y)$$

And when it takes the second word $x^{<2>}$, it also gets information from the previous step using the non-activated neurons.

$$a^{<2>} = tanh(W_{aa}a^{<1>} + W_{ax}x^{<2>} + b_a)$$
$$\hat{y}^{<2>} = \sigma(W_{ya}a^{<2>} + b_y)$$

For the tth word:

$$a^{<t>} = tanh(W_{aa}a^{<t-1>} + W_{ax}x^{<t>} + b_a)$$
$$\hat{y}^{<t>} = \sigma(W_{ya}a^{<t>} + b_y)$$

The information flows from one step to the next with a repetitive structure until the last time step input $x^{<T_x>}$ and then it outputs $\hat{y}^{<T_y>}$. In this example, $T_x = T_y$. The architecture changes when T_x and T_y are not the same. The model shares parameters, $W_{ya}, W_{aa}, W_{ax}, b_a, b_y$, for all time steps of the input.

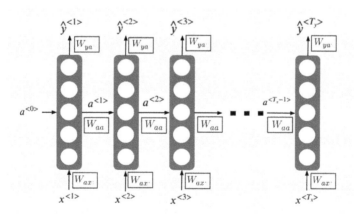

Calculate the loss function:

$$L^{<t>}(\hat{y}^{<t>}) = -y^{<t>}log(\hat{y}^{<t>}) - (1 - y^{<t>})log(1 - \hat{y}^{<t>})$$
$$L(\hat{y}, y) = \Sigma_{t=1}^{T_y}L^{<t>}(\hat{y}, y)$$

The above defines the forward process. Same as before, the backward propagation computes the gradient descent for the parameters by the chain rule for differentiation.

In this RNN structure, the information only flows from the left to the right. So at any position, it only uses data from earlier in the sequence to make a prediction. It does not work when predicting the current word needs information from later words. For example, consider the following two sentences:

1. Do you like April Kepner in Grey's Anatomy?
2. Do you like April in Los Angeles? It is not too hot.

Given just the first three words is not enough to know if the word "April" is part of a person's name. It is a person's name in 1 but not 2. The two sentences have the same first three words. In this case, we need a model that allows the information to flow in both directions. A bidirectional RNN takes data from both earlier and later in the sequence. The disadvantage is that it needs the entire word sequence to predict at any position. For a speech recognition application that requires capturing the speech in real-time, we need a more complex method called the attention model. We will not get into those models here. Deep Learning with R (Chollet and Allaire, 2018) provides a high-level introduction of bidirectional RNN with applicable codes. It teaches both intuition and practical, computational usage of deep learning models. For python users, refer to Deep Learning with Python (Chollet, 2017). A standard text with heavy mathematics is Deep Learning (Goodfellow et al., 2016).

12.3.2 Long Short Term Memory

The sequence in RNN can be very long, which leads to the vanishing gradient problem even when the RNN network is not deep. Think about the following examples:

1. The **girl** walked away, sat down in the shade of a tree, and began to read a new book which **she** bought the day before.

2.   The **boy** walked away, sat down in the shade of a tree, and began to read a new book which **he** bought the day before.

For sentence 1, you need to use "she" in the adjective clause after "which" because it is a girl. For sentence 2, you need to use "he" because it is a boy. This is a long-term dependency example where the information at the beginning can affect what needs to come much later in the sentence. RNN needs to forward propagate information from left to right and then backpropagate from right to left. It can be difficult for the error associated with the later sequence to affect the optimization earlier. So in practice, it means the model might fail to do the task mentioned above. People came up with different methods to mitigate this issue, such as the Greater Recurrent Units (GRU) (Chung et al., 2014) and Long Short Term Memory Units (LSTM) (Hochreiter and Schmidhuber, 1997). The goal is to help the model memorize information in the earlier sequence. We are going to walk through LSTM step by step.

The first step of LSTM is to decide what information to **forget**. This decision is made by "forget gate," a sigmoid function (Γ_f). It looks at $a^{<t-1>}$ and x^t and outputs a number between 0 and 1 for each number in the cell state c^{t-1}. A value 1 means "completely remember the state," while 0 means "completely forget the state."

$$\Gamma_f = \sigma(W_f[a^{<t-1>}, x^{<t>}] + b_f)$$

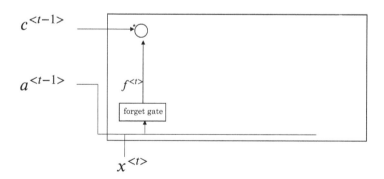

The next step is to decide what new information we're going to add to the cell state. This step includes two parts:

1. input gate (Γ_u): a sigmoid function that decides how much we want to update
2. a vector of new candidate value ($\tilde{c}^{<t>}$)

$$\Gamma_u = \sigma(W_u[a^{<t-1>}, x^{<t>}] + b_u)$$
$$\tilde{c}^{<t>} = tanh(W_c[a^{<t-1>}, x^{<t>}] + b_c)$$

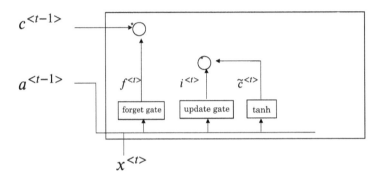

The multiplication of the above two parts $\Gamma_u * \tilde{c}^{<t>}$ is the new candidate scaled by the input gate. We then combine the results we get so far to get new cell state $c^{<t>}$.

$$c^{<t>} = \Gamma_u * \tilde{c}^{<t>} + \Gamma_f * c^{<t-1>}$$

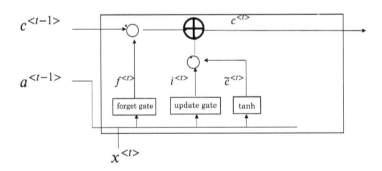

Finally, we need to decide what we are going to output. The output is a filtered version of the new cell state $c^{<t>}$.

$$\Gamma_0 = \sigma(W_0[a^{<t-1>}, x^{<t>}] + b_0)$$
$$a^{<t>} = \Gamma_0 * tanh(c^{<t>})$$

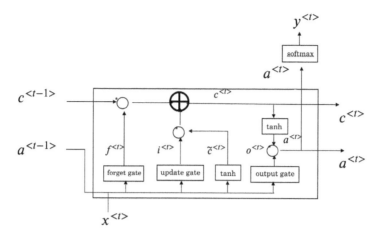

12.3.3 Word Embedding

So far, we have been using one-hot encoding to represent the words. This representation is sparse and doesn't capture the relationship between the words. For example, if your model learns from the training data that the word after "pumpkin" in the first sentence is "pie," can it fill the second sentence's blank with "pie"?

1. [training data] My favorite Christmas dessert is pumpkin pie.
2. [testing data] My favorite Christmas dessert is apple _____.

The algorithm cannot learn the relationship between "pumpkin" and "apple" by the distance between the one-hot encoding for the two words. If we can find a way to create features to represent the words, we may teach the model to learn that pumpkin and apple is food. And the distance between the feature representations of these two words is closer than, for example, "apple" and "nurse." So when the model sees "apple" in a new sentence and needs to predict the word after, it is more likely to choose "pie" since it sees "pumpkin pie" in the training data. The idea of word embedding is to learn a set of features to represent words. For example, you can score each word in the dictionary according to a group of features like this:

	Nurse (5391)	Engineer (3853)	Apple (456)	Pumpkin (7157)	Male (2456)	Female (1632)
Food	0.01	0.02	0.98	0.96	- 0.01	0.01
Occupation	0.95	0.92	0.09	0.01	0.02	0.01
Transportation	-0.02	0.01	0.05	0.03	0.02	0.01
Gender	0.01	0.01	0.05	0.03	-1	1
People	0.85	0.82	0.01	0.01	0.92	0.92
.						
.						
.						

The word "male" has a score of -1 for the "gender" feature, "female" has a score of 1. Both "Apple" and "pumpkin" have a high score for the "food" feature and much lower scores for the rest. You can set the number of features to learn, usually more than what we list in the above figure. If you use 200 features to represent the words, then the learned embedding for each word is a vector with a length of 200.

For language-related applications, text embedding is the most critical step. It converts raw text into a meaningful vector representation. Once we have a vector representation, it is easy to calculate typical numerical metrics such as cosine similarity. There are many pre-trained text embeddings available for us to use. We will briefly introduce some of these popular embeddings.

The first widely used embedding is word2vec. It was first introduced in 2013 and was trained by a large collection of text in an unsupervised way. Training the word2vec embedding vector uses bag-of-words or skip-gram. In the bag-of-words architecture, the model predicts the current word based on a window of surrounding context words. In skip-gram architecture, the model uses the current word to predict the surrounding window of context words. There are pre-trained word2vec embeddings based on a large amount of text (such as wiki pages, news reports, etc.) for general applications.

GloVe (Global Vectors) embedding is an extension of word2vec and performs better. It uses a unique version of the square loss function. However, words are composite of meaningful components such as radicals.

"eat" and "eaten" are different forms of the same word. Both word2vec and GloVe use word-level information, and they treat each word uniquely based on its context.

The fastText embedding is introduced to use the word's internal structure to make the process more efficient. It uses morphological information to extend the skip-gram model. New words that are not in the training data can be repressed well. It also supports 150+ different languages. The above-mentioned embeddings (word2vec, GloVe, and fastText) do not consider the words' context (i.e., the same word has the same embedding vector). However, the same word may have different meanings in a different context.

More recently transformer based networks, such as BERT (Bidirectional Encoder Representations from Transformers), were introduced to add context-level information in text-related applications. These models use a new mechanism, attention, to read a sequence simultaneously instead of the one-input-at-a-time process of RNNs. These networks combine positional embeddings along with embeddings for each token in the sequence giving it the ability to differentiate different uses of the same word based on surrounding words.

12.3.4 Sentiment Analysis Using RNN

In this section, we will walk through an example of text sentiment analysis using RNN. Refer to section 4.3 to set up an account, create a notebook (R or Python) and start a cluster. Refer to section 12.1.7 for package installation.

We will use the IMDB movie review data. It is one of the most used datasets for text-related machine learning methods. The datasets' inputs are movie reviews published at IMDB in its raw text format, and the output is a binary sentiment indicator("1" for positive and "0" for negative) created through human evaluation. The training and testing data have 25,000 records each. Each review varies in length.

12.3.4.1 Data Preprocessing

Machine learning algorithms cannot deal with raw text, and we have to convert text into numbers before feeding it into an algorithm.

Tokenization is one way to convert text data into a numerical representation. For example, suppose we have 500 unique words for all reviews in the training dataset. We can label each word by the rank (i.e., from 1 to 500) of their frequency in the training data. Then each word is replaced by an integer between 1 to 500. This way, we can map each movie review from its raw text format to a sequence of integers.

As reviews can have different lengths, sequences of integers will have different sizes too. So another important step is to make sure each input has the same length by padding or truncating. For example, we can set a length of 50 words, and for any reviews less than 50 words, we can pad 0 to make it 50 in length; and for reviews with more than 50 words, we can truncate the sequence to 50 by keeping only the first 50 words. After padding and truncating, we have a typical data frame, each row is an observation, and each column is a feature. The number of features is the number of words designed for each review (i.e., 50 in this example).

After tokenization, the numerical input is just a naive mapping to the original words, and the integers do not have their usual numerical meanings. We need to use embedding to convert these categorical integers to more meaningful representations. The word embedding captures the inherited relationship of words and dramatically reduces the input dimension (see section 12.3.3). The dimension is a vector space representing the entire vocabulary. It can be 128 or 256, and the vector space dimension is the same when the vocabulary changes. It has a lower dimension, and each vector is filled with real numbers. The embedding vectors can be learned from the training data, or we can use pre-trained embedding models. There are many pre-trained embeddings for us to use, such as Word2Vec, BIRD.

12.3.4.2   R Code for IMDB Dataset

The IMDB dataset is preloaded for `keras` and we can call `dataset_imdb()` to load a partially pre-processed dataset into a data frame. We can define a few parameters in that function. `num_words` is the number of words in each review to be used. All the unique words are ranked by their frequency counts in the training dataset.

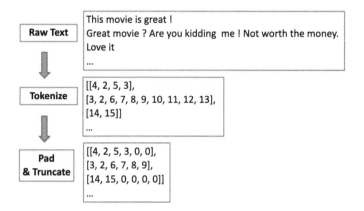

The `dataset_imdb()` function keeps the top `num_words` words and replaces other words with a default value of 2, and using integers to represent text (i.e., top frequency word will be replaced by 3 and 0, 1, 2 are reserved for "padding," "start of the sequence," and "unknown.").

```
# Load `keras` package
library(keras)

# consider only the top 10,000 words in the dataset
max_unique_word <- 2500
# cut off reviews after 100 words
max_review_len <- 100
```

Now we load the IMDB dataset, and we can check the structure of the loaded object by using `str()` command.

```
my_imdb <- dataset_imdb(num_words = max_unique_word)
str(my_imdb)
```

```
Downloading data from
https://storage.googleapis.com/tensorflow/tf-keras-datasets/imdb.npz

    8192/17464789 [..............................] - ETA: 0s
  811008/17464789 [>.............................] - ETA: 1s
 4202496/17464789 [======>.......................] - ETA: 0s
11476992/17464789 [===================>..........] - ETA: 0s
17465344/17464789 [==============================] - 0s 0us/step
List of 2
 $ train:List of 2
  ..$ x:List of 25000
  .. ..$ : int [1:218] 1 14 22 16 43 530 973 1622 1385 65 ...
  .. ..$ : int [1:189] 1 194 1153 194 2 78 228 5 6 1463 ...

*** skipped some output ***
```

```
x_train <- my_imdb$train$x
y_train <- my_imdb$train$y
x_test  <- my_imdb$test$x
y_test  <- my_imdb$test$y
```

Next, we do the padding and truncating process.

```
x_train <- pad_sequences(x_train, maxlen = max_review_len)
x_test <- pad_sequences(x_test, maxlen = max_review_len)
```

The x_train and x_test are numerical data frames ready to be used for recurrent neural network models.

Simple Recurrent Neural Network

Like DNN and CNN models we trained in the past, RNN models are relatively easy to train using keras after the pre-processing stage.

In the following example, we use `layer_embedding()` to fit an embedding layer based on the training dataset, which has two parameters: `input_dim` (the number of unique words) and `output_dim` (the length of dense vectors). Then, we add a simple RNN layer by calling `layer_simple_rnn()` and followed by a dense layer `layer_dense()` to connect to the response binary variable.

```
rnn_model <- keras_model_sequential()
rnn_model %>%
  layer_embedding(input_dim = max_unique_word, output_dim = 128) %>%
  layer_simple_rnn(units = 64, dropout = 0.2,
                   recurrent_dropout = 0.2) %>%
  layer_dense(units = 1, activation = 'sigmoid')
```

We compile the RNN model by defining the loss function, optimizer to use, and metrics to track the same way as DNN and CNN models.

```
rnn_model %>% compile(
  loss = 'binary_crossentropy',
  optimizer = 'adam',
  metrics = c('accuracy')
)
```

Let us define a few more variables before fitting the model: `batch_size`, `epochs`, and `validation_split`. These variables have the same meaning as DNN and CNN models we see in the past.

```
batch_size = 128
epochs = 5
validation_split = 0.2

rnn_history <- rnn_model %>% fit(
  x_train, y_train,
```

```
  batch_size = batch_size,
  epochs = epochs,
  validation_split = validation_split
)
```

```
plot(rnn_history)
```

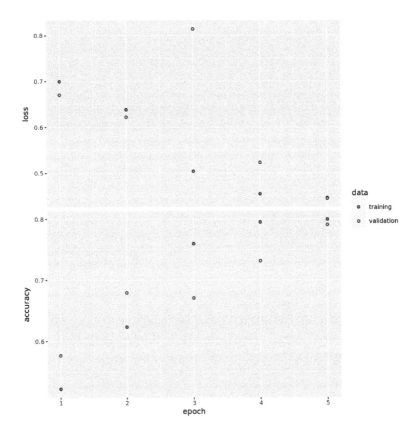

```
rnn_model %>%
    evaluate(x_test, y_test)
```

```
##       loss  accuracy
## 0.5441073 0.7216800
```

LSTM RNN Model

A simple RNN layer is a good starting point for learning RNN, but the performance is usually not that good because these long-term dependencies are impossible to learn due to vanishing gradient. Long Short Term Memory RNN model (LSTM) can carry useful information from the earlier words to later words. In keras, it is easy to replace a simple RNN layer with an LSTM layer by using layer_lstm().

```
lstm_model <- keras_model_sequential()

lstm_model %>%
  layer_embedding(input_dim = max_unique_word, output_dim = 128) %>%
  layer_lstm(units = 64, dropout = 0.2, recurrent_dropout = 0.2) %>%
  layer_dense(units = 1, activation = 'sigmoid')

lstm_model %>% compile(
  loss = 'binary_crossentropy',
  optimizer = 'adam',
  metrics = c('accuracy')
)

batch_size = 128
epochs = 5
validation_split = 0.2

lstm_history <- lstm_model %>% fit(
  x_train, y_train,
  batch_size = batch_size,
  epochs = epochs,
  validation_split = validation_split
)
```

```
plot(lstm_history)
```

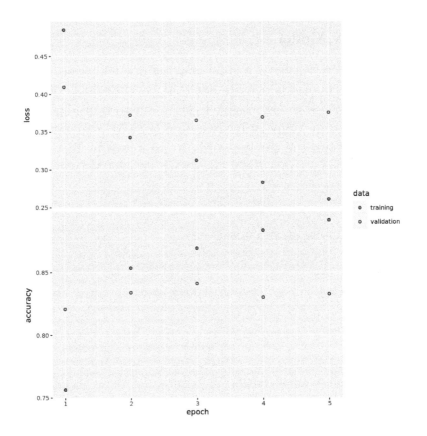

```
lstm_model %>%
    evaluate(x_test, y_test)
```

```
##       loss accuracy
## 0.361364 0.844080
```

This simple example shows that LSTM's performance has improved dramatically from the simple RNN model. The computation time for LSTM is roughly doubled when compared with the simple RNN model for this small dataset.

A

Handling Large Local Data

When the data is too large to fit in a computer's memory, we can use some big data analytics engine like Spark on a cloud platform (see Chapter 4). However, even when the data can fit in the memory, there may be a situation where it is slow to read and manipulate due to a relatively large size. Some R packages can make the process faster with the cost of familiarity, especially for data wrangling. But it avoids the hurdle of setting up Spark cluster and working in an unfamiliar environment. This section presents some of the alternative R packages to read, write and wrangle a data set that is relatively large but not too big to fit in the memory.

Load the R packages first:

```
# install packages from CRAN if you haven't
library(readr)
library(data.table)
```

A.1 readr

You must be familiar with `read.csv()`, `read.table()` and `write.csv()` in base R. Here we will introduce a more efficient package for reading and writing data: `readr` package. The corresponding functions are `read_csv()`, `read_table()` and `write_csv()`. The commands look quite similar, but `readr` is different in the following respects:

1. It is 10x faster. The trick is that `readr` uses C++ to process the data quickly.

DOI: 10.1201/9781351132916-A

2. It doesn't change the column names. The names can start with a number and "." will not be substituted to "_". For example:

```
read_csv("2015,2016,2017
1,2,3
4,5,6")
```

```
## Rows: 2 Columns: 3
## -- Column specification ------------------------------
## Delimiter: ","
## dbl (3): 2015, 2016, 2017
##
## i Use `spec()` to retrieve the full column specification for
   this data.
## i Specify the column types or set `show_col_types = FALSE` to
   quiet this message.

## # A tibble: 2 x 3
##    `2015` `2016` `2017`
##     <dbl>  <dbl>  <dbl>
## 1      1      2      3
## 2      4      5      6
```

1. readr functions do not convert strings to factors by default, are able to parse dates and times and can automatically determine the data types in each column.

2. The killing character, in my opinion, is that readr provides **progress bar**. What makes you feel worse than waiting is not knowing how long you have to wait.

The major functions of readr is to turn flat files into data frames:

- `read_csv()`: reads comma delimited files
- `read_csv2()`: reads semicolon separated files (common in countries where, is used as the decimal place)
- `read_tsv()`: reads tab delimited files
- `read_delim()`: reads in files with any delimiter
- `read_fwf()`: reads fixed width files. You can specify fields either by their widths with `fwf_widths()` or their position with `fwf_positions()`

- `read_table()`: reads a common variation of fixed width files where columns are separated by white space
- `read_log()`: reads Apache style log files

The good thing is that those functions have similar syntax. Once you learn one, the others become easy. Here we will focus on `read_csv()`.

The most important information for `read_csv()` is the path to your data:

```
sim.dat <- read_csv("http://bit.ly/2P5gTw4")
head(sim.dat)
```

```
# A tibble: 6 x 19
    age gender income house store_exp online_exp store_trans online_trans    Q1
  <int> <chr>   <dbl> <chr>     <dbl>      <dbl>       <int>        <int> <int>
1    57 Female 1.21e5 Yes        529.       304.           2            2     4
2    63 Female 1.22e5 Yes        478.       110.           4            2     4
3    59 Male   1.14e5 Yes        491.       279.           7            2     5
4    60 Male   1.14e5 Yes        348.       142.          10            2     5
5    51 Male   1.24e5 Yes        380.       112.           4            4     4
6    59 Male   1.08e5 Yes        338.       196.           4            5     4
# ... with 10 more variables: Q2 <int>, Q3 <int>, Q4 <int>, Q5 <int>, Q6 <int>,
#   Q7 <int>, Q8 <int>, Q9 <int>, Q10 <int>, segment <chr>
```

The function reads the file to R as a `tibble`. You can consider `tibble` as next iteration of the data frame. They are different with data frame for the following aspects:

- It never changes an input's type (i.e., no more `stringsAsFactors = FALSE`!)
- It never adjusts the names of variables
- It has a refined print method that shows only the first 10 rows and all the columns that fit on the screen. You can also control the default print behavior by setting options.

Refer to `http://r4ds.had.co.nz/tibbles.html` for more information about "tibble."

When you run `read_csv()` it prints out a column specification that gives the name and type of each column. To better understanding how `readr` works, it is helpful to type in some baby data set and check the results:

```
dat <- read_csv("2015,2016,2017
100,200,300
canola,soybean,corn")
```

```
print(dat)
```

```
## # A tibble: 2 x 3
##   `2015` `2016`  `2017`
##   <chr>  <chr>   <chr>
## 1 100    200     300
## 2 canola soybean corn
```

You can also add comments on the top and tell R to skip those lines:

```
dat <- read_csv("# I will never let you know that
          # my favorite food is carrot
          Date,Food,Mood
          Monday,carrot,happy
          Tuesday,carrot,happy
          Wednesday,carrot,happy
          Thursday,carrot,happy
          Friday,carrot,happy
          Saturday,carrot,extremely happy
          Sunday,carrot,extremely happy",
          skip = 2)
```

```
print(dat)
```

```
## # A tibble: 7 x 3
##    Date      Food    Mood
##    <chr>     <chr>   <chr>
## 1 Monday    carrot happy
## 2 Tuesday   carrot happy
## 3 Wednesday carrot happy
## 4 Thursday  carrot happy
## 5 Friday    carrot happy
## 6 Saturday  carrot extremely happy
## 7 Sunday    carrot extremely happy
```

If you don't have column names, set col_names = FALSE then R will assign names "x1","x2"... to the columns:

```
dat <- read_csv("Saturday,carrot,extremely happy
          Sunday,carrot,extremely happy", col_names = FALSE)
```

```
print(dat)
```

```
## # A tibble: 2 x 3
##   X1       X2      X3
##   <chr>    <chr>   <chr>
## 1 Saturday carrot extremely happy
## 2 Sunday   carrot extremely happy
```

You can also pass `col_names` a character vector which will be used as the column names. Try to replace `col_names=FALSE` with `col_names=c("Date","Food","Mood")` and see what happen.

As mentioned before, you can use `read_csv2()` to read semicolon separated files:

```
dat <- read_csv2("Saturday; carrot; extremely happy \n
             Sunday; carrot; extremely happy", col_names = FALSE)
print(dat)
```

```
## # A tibble: 2 x 3
##   X1       X2      X3
##   <chr>    <chr>   <chr>
## 1 Saturday carrot extremely happy
## 2 Sunday   carrot extremely happy
```

Here "\n" is a convenient shortcut for adding a new line.
You can use `read_tsv()` to read tab delimited files:

```
dat <- read_tsv("every\tman\tis\ta\tpoet\twhen\the\tis\tin\tlove\n",
   col_names = FALSE)
```

```
print(dat)
```

```
## # A tibble: 1 x 10
##    X1    X2    X3    X4    X5    X6    X7    X8    X9
##    <chr> <chr> <chr> <chr> <chr> <chr> <chr> <chr> <chr>
## 1 every man   is    a     poet  when  he    is    in
## # ... with 1 more variable: X10 <chr>
```

Or more generally, you can use `read_delim()` and assign separating character:

```
dat <- read_delim("THE|UNBEARABLE|RANDOMNESS|OF|LIFE\n",
    delim = "|", col_names = FALSE)
```

```
print(dat)
```

```
## # A tibble: 1 x 5
##    X1    X2         X3         X4    X5
##    <chr> <chr>      <chr>      <chr> <chr>
## 1 THE    UNBEARABLE RANDOMNESS OF    LIFE
```

Another situation you will often run into is the missing value. In marketing survey, people like to use "99" to represent missing. You can tell R to set all observation with value "99" as missing when you read the data:

```
dat <- read_csv("Q1,Q2,Q3
               5, 4,99",
               na = "99")
```

```
print(dat)
```

```
## # A tibble: 1 x 3
##      Q1    Q2 Q3
##   <dbl> <dbl> <lgl>
## 1     5     4 NA
```

For writing data back to disk, you can use `write_csv()` and `write_tsv()`. The following two characters of the two functions increase the chances of the output file being read back in correctly:

- Encode strings in UTF-8
- Save dates and date-times in ISO8601 format so they are easily parsed elsewhere

For example:

```
write_csv(sim.dat, "sim_dat.csv")
```

For other data types, you can use the following packages:

- `Haven`: SPSS, Stata and SAS data
- `Readxl` and `xlsx`: excel data(.xls and .xlsx)
- `DBI`: given data base, such as RMySQL, RSQLite and RPost-greSQL, read data directly from the database using SQL

Some other useful materials:

- For getting data from the internet, you can refer to the book "XML and Web Technologies for Data Sciences with R."

- R data import/export manual[1]
- `rio` package:`https://github.com/leeper/rio`

---

[1]`https://cran.r-project.org/doc/manuals/r-release/R-data.html#Acknowledgeme`
`nts`

A.2   `data.table`— Enhanced `data.frame`

What is `data.table`? It is an R package that provides an enhanced version of `data.frame`. The most used object in R is `data frame`. Before we move on, let's briefly review some basic characters and manipulations of data.frame:

- It is a set of rows and columns.
- Each row is of the same length and data type
- Every column is of the same length but can be of differing data types
- It has characteristics of both a matrix and a list
- It uses `[]` to subset data

We will use the clothes customer data to illustrate. There are two dimensions in `[]`. The first one indicates the row and second one indicates column. It uses a comma to separate them.

```
# read data
sim.dat <- readr::read_csv("http://bit.ly/2P5gTw4")
```

```
# subset the first two rows
sim.dat[1:2, ]
# subset the first two rows and column 3 and 5
sim.dat[1:2, c(3, 5)]
# get all rows with age>70
sim.dat[sim.dat$age > 70, ]
# get rows with age> 60 and gender is Male select column 3 and 4
sim.dat[sim.dat$age > 68 & sim.dat$gender == "Male", 3:4]
```

Remember that there are usually different ways to conduct the same manipulation. For example, the following code presents three ways to calculate an average number of online transactions for male and female:

```
tapply(sim.dat$online_trans, sim.dat$gender, mean)

aggregate(online_trans ~ gender, data = sim.dat, mean)

sim.dat %>%
  group_by(gender) %>%
  summarise(Avg_online_trans = mean(online_trans))
```

There is no gold standard to choose a specific function to
manipulate data. The goal is to solve the real problem, not the
tool itself. So just use whatever tool that is convenient for you.

The way to use [] is straightforward. But the manipulations are
limited. If you need more complicated data reshaping or aggregation,
there are other packages to use such as dplyr, reshape2, tidyr etc.
But the usage of those packages are not as straightforward as [].
You often need to change functions. Keeping related operations
together, such as subset, group, update, join etc, will allow for:

- concise, consistent and readable syntax irrespective of the set
  of operations you would like to perform to achieve your end
  goal
- performing data manipulation fluidly without the cognitive
  burden of having to change among different functions
- by knowing precisely the data required for each operation, you
  can automatically optimize operations effectively

data.table is the package for that. If you are not familiar with
other data manipulating packages and are interested in reducing
programming time tremendously, then this package is for you.

Other than extending the function of [], data.table has the
following advantages:

- Offers fast import, subset, grouping, update, and joins for large
  data files
- It is easy to turn data frame to data table
- Can behave just like a data frame

You need to install and load the package:

Use data.table() to convert the existing data frame sim.dat to data table:

```
dt <- data.table(sim.dat)
class(dt)
```

```
## [1] "data.table" "data.frame"
```

Calculate mean for counts of online transactions:

```
dt[, mean(online_trans)]
```

```
## [1] 13.55
```

You can't do the same thing using data frame:

```
sim.dat[,mean(online_trans)]
```

```
Error in mean(online_trans) : object 'online_trans' not found
```

If you want to calculate mean by group as before, set "by =" argument:

```
dt[ , mean(online_trans), by = gender]
```

```
##     gender   V1
## 1: Female 15.38
## 2:   Male 11.26
```

You can group by more than one variables. For example, group by "gender" and "house":

```
dt[ , mean(online_trans), by = .(gender, house)]
```

```
##     gender house      V1
## 1: Female    Yes 11.312
## 2:   Male    Yes  8.772
## 3: Female     No 19.146
## 4:   Male     No 16.486
```

Assign column names for aggregated variables:

```
dt[ , .(avg = mean(online_trans)), by = .(gender, house)]
```

```
##     gender house     avg
## 1: Female    Yes 11.312
## 2:   Male    Yes  8.772
## 3: Female     No 19.146
## 4:   Male     No 16.486
```

data.table can accomplish all operations that aggregate() and tapply()can do for data frame.

- General setting of data.table

Different from data frame, there are three arguments for data table:

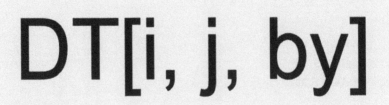

It is analogous to SQL. You don't have to know SQL to learn data table. But experience with SQL will help you understand data table. In SQL, you select column j (use command SELECT) for row i (using command WHERE). GROUP BY in SQL will assign the variable to group the observations.

R	**:**	**i**	**j**	**by**
SQL	**:**	**WHERE**	**SELECT**	**GROUP BY**

Let's review our previous code:

```
dt[ , mean(online_trans), by = gender]
```

The code above is equal to the following SQL:

```
SELECT
    gender,
    avg(online_trans)
FROM
    sim.dat
GROUP BY
    gender
```

R code:

```
dt[ , .(avg = mean(online_trans)), by = .(gender, house)]
```

is equal to SQL:

```
SELECT
    gender,
    house,
    avg(online_trans) AS avg
FROM
    sim.dat
GROUP BY
    gender,
    house
```

R code:

```
dt[ age < 40, .(avg = mean(online_trans)), by = .(gender, house)]
```

is equal to SQL:

```
SELECT
    gender,
    house,
    avg(online_trans) AS avg
FROM
    sim.dat
WHERE
    age < 40
GROUP BY
    gender,
    house
```

You can see the analogy between `data.table` and `SQL`. Now let's focus on operations in data table.

- select row

```
# select rows with age<20 and income > 80000
dt[age < 20 & income > 80000]
```

```
##     age gender income house store_exp online_exp
## 1:  19 Female  83535    No     227.7        1491
## 2:  18 Female  89416   Yes     209.5        1926
## 3:  19 Female  92813    No     186.7        1042
##     store_trans online_trans Q1 Q2 Q3 Q4 Q5 Q6 Q7 Q8 Q9
## 1:            1           22  2  1  1  2  4  1  4  2  4
## 2:            3           28  2  1  1  1  4  1  4  2  4
## 3:            2           18  3  1  1  2  4  1  4  3  4
##     Q10 segment
## 1:    1   Style
## 2:    1   Style
## 3:    1   Style
```

```
# select the first two rows
dt[1:2]
```

```
##     age gender income house store_exp online_exp
## 1:  57 Female 120963   Yes     529.1      303.5
## 2:  63 Female 122008   Yes     478.0      109.5
##     store_trans online_trans Q1 Q2 Q3 Q4 Q5 Q6 Q7 Q8 Q9
## 1:            2            2  4  2  1  2  1  4  1  4  2
## 2:            4            2  4  1  1  2  1  4  1  4  1
##     Q10 segment
## 1:    4   Price
## 2:    4   Price
```

- select column

Selecting columns in `data.table` don't need `$`:

```
# select column "age" but return it as a vector
# the argument for row is empty so the result
# will return all observations
ans <- dt[, age]
head(ans)
```

```
## [1] 57 63 59 60 51 59
```

To return `data.table` object, put column names in `list()`:

```
# Select age and online_exp columns
# and return as a data.table instead
ans <- dt[, list(age, online_exp)]
head(ans)
```

Or you can also put column names in `.()`:

```
ans <- dt[, .(age, online_exp)]
```

To select all columns from "age" to "income":

```
ans <- dt[, age:income, with = FALSE]
head(ans,2)
```

```
##     age gender income
## 1:   57 Female 120963
## 2:   63 Female 122008
```

Delete columns using - or !:

```
# delete columns from  age to online_exp
ans <- dt[, -(age:online_exp), with = FALSE]
ans <- dt[, !(age:online_exp), with = FALSE]
```

- tabulation

  In data table, .N means to count.

```
# row count
dt[, .N]
```

```
## [1] 1000
```

If you assign the group variable, then it will count by groups:

```
# counts by gender
dt[, .N, by= gender]
```

```
##     gender   N
## 1: Female 554
## 2:   Male 446
```

```
# for those younger than 30, count by gender
 dt[age < 30, .(count=.N), by= gender]
```

```
##     gender count
## 1: Female   292
## 2:   Male    86
```

Order table:

```
# get records with the highest 5 online expense:
head(dt[order(-online_exp)],5)
```

```
   age gender   income house store_exp online_exp store_trans ...
1:  40 Female 217599.7    No  7023.684   9479.442          10
2:  41 Female       NA   Yes  3786.740   8638.239          14
3:  36   Male 228550.1   Yes  3279.621   8220.555           8
4:  31 Female 159508.1   Yes  5177.081   8005.932          11
5:  43 Female 190407.4   Yes  4694.922   7875.562           6
...
```

Since data table keep some characters of data frame, they share some operations:

```
dt[order(-online_exp)][1:5]
```

You can also order the table by more than one variable. The following code will order the table by gender, then order within gender by online_exp:

```
dt[order(gender, -online_exp)][1:5]
```

- Use fread() to import dat

Other than read.csv in base R, we have introduced "read_csv" in "readr". read_csv is much faster and will provide progress bar which makes user feel much better (at least make me feel better). fread() in data.table further increase the efficiency of reading data. The following are three examples of reading the same data file topic.csv. The file includes text data scraped from an agriculture forum with 209670 rows and 6 columns:

```
system.time(topic <- read.csv("http://bit.ly/2zam5ny"))
```

```
   user  system elapsed
  3.561   0.051   4.888
```

```
system.time(topic <- readr::read_csv("http://bit.ly/2zam5ny"))
```

```
   user  system elapsed
  0.409   0.032   2.233
```

```
system.time(topic <- data.table::fread("http://bit.ly/2zam5ny"))
```

```
   user  system elapsed
  0.276   0.096   1.117
```

It is clear that `read_csv()` is much faster than `read.csv()`. `fread()` is a little faster than `read_csv()`. As the size increasing, the difference will become for significant. Note that `fread()` will read file as `data.table` by default.

B

R Code for Data Simulation

B.1 Customer Data for Clothing Company

The simulation is not very straightforward and we will break it into three parts:

1. Define data structure: variable names, variable distribution, customer segment names, segment size
2. Variable distribution parameters: mean and variance
3. Iterate across segments and variables. Simulate data according to specific parameters assigned

By organizing code this way, it makes easy for us to change specific parts of the simulation. For example, if we want to change the distribution of one variable, we can just change the corresponding part of the code.

Here is code to define data structure:

```r
# set a random number seed to
# make the process repeatable
set.seed(12345)
# define the number of observations
ncust <- 1000
# create a data frmae for simulated data
seg_dat <- data.frame(id = as.factor(c(1:ncust)))
# assign the variable names
vars <- c("age", "gender", "income", "house", "store_exp",
    "online_exp", "store_trans", "online_trans")
# assign distribution for each variable
```

DOI: 10.1201/9781351132916-B

```r
vartype <- c("norm", "binom", "norm", "binom", "norm", "norm",
    "pois", "pois")
# names of 4 segments
group_name <- c("Price", "Conspicuous", "Quality", "Style")
# size of each segments
group_size <- c(250, 200, 200, 350)
```

The next step is to define variable distribution parameters. There are 4 segments of customers and 8 parameters. Different segments correspond to different parameters. Let's store the parameters in a 4×8 matrix:

```r
# matrix for mean
mus <- matrix( c(
  # Price
  60, 0.5, 120000,0.9, 500,200,5,2,
  # Conspicuous
  40, 0.7, 200000,0.9, 5000,5000,10,10,
  # Quality
  36, 0.5, 70000, 0.4, 300, 2000,2,15,
  # Style
  25, 0.2, 90000, 0.2, 200, 2000,2,20),
  ncol=length(vars), byrow=TRUE)
```

```r
# matrix for variance
sds<- matrix( c(
  # Price
  3,NA,8000,NA,100,50,NA,NA,
  # Conspicuous
  5,NA,50000,NA,1000,1500,NA,NA,
  # Quality
  7,NA,10000,NA,50,200,NA,NA,
  # Style
```

```
   2,NA,5000,NA,10,500,NA,NA),
   ncol=length(vars), byrow=TRUE)
```

Now we are ready to simulate data using the parameters defined above:

```
# simulate non-survey data
sim.dat <- NULL
set.seed(2016)
# loop on customer segment (i)
for (i in seq_along(group_name)) {

    # add this line in order to moniter the process
    cat(i, group_name[i], "\n")

    # create an empty matrix to store relevent data
    seg <- data.frame(matrix(NA, nrow = group_size[i],
    ncol = length(vars)))

    # Simulate data within segment i
    for (j in seq_along(vars)) {

        # loop on every variable (j)
        if (vartype[j] == "norm") {
            # simulate normal distribution
            seg[, j] <- rnorm(group_size[i], mean = mus[i,
                j], sd = sds[i, j])
        } else if (vartype[j] == "pois") {
            # simulate poisson distribution
            seg[, j] <- rpois(group_size[i], lambda = mus[i,
                j])
        } else if (vartype[j] == "binom") {
            # simulate binomial distribution
            seg[, j] <- rbinom(group_size[i], size = 1,
                prob = mus[i, j])
```

```
        } else {
            # if the distribution name is not one of the above, stop
            # and return a message
            stop("Don't have type:", vartype[j])
        }
    }
    sim.dat <- rbind(sim.dat, seg)
}
```

Now let's edit the data we just simulated a little by adding tags to 0/1 binomial variables:

```
# assign variable names
names(sim.dat) <- vars
# assign factor levels to segment variable
sim.dat$segment <- factor(rep(group_name, times = group_size))
# recode gender and house variable
sim.dat$gender <- factor(sim.dat$gender, labels = c("Female",
    "Male"))
sim.dat$house <- factor(sim.dat$house, labels = c("No",
    "Yes"))
# store_trans and online_trans are at least 1
sim.dat$store_trans <- sim.dat$store_trans + 1
sim.dat$online_trans <- sim.dat$online_trans + 1
# age is integer
sim.dat$age <- floor(sim.dat$age)
```

In the real world, the data always includes some noise such as missing, wrong imputation. So we will add some noise to the data:

```
# add missing values
idxm <- as.logical(rbinom(ncust, size = 1, prob = sim.dat$age/200))
sim.dat$income[idxm] <- NA
# add wrong imputations and outliers
```

```r
set.seed(123)
idx <- sample(1:ncust, 5)
sim.dat$age[idx[1]] <- 300
sim.dat$store_exp[idx[2]] <- -500
sim.dat$store_exp[idx[3:5]] <- c(50000, 30000, 30000)
```

So far we have created part of the data. You can check it using `summary(sim.dat)`. Next, we will move on to simulate survey data.

```r
# number of survey questions
nq <- 10

# mean matrix for different segments
mus2 <- matrix( c( 5,2,1,3,1,4,1,4,2,4, # Price
  1,4,5,4,4,4,4,1,4,2, # Conspicuous
  5,2,3,4,3,2,4,2,3,3, # Quality
  3,1,1,2,4,1,5,3,4,2), # Style
ncol=nq, byrow=TRUE)

# assume the variance is 0.2 for all
sd2 <- 0.2
sim.dat2 <- NULL
set.seed(1000)
# loop for customer segment (i)
for (i in seq_along(group_name)) {
    # the following line is used for checking the
    # progress cat (i, group_name[i],'\n') create an
    # empty data frame to store data
    seg <- data.frame(matrix(NA, nrow = group_size[i],
        ncol = nq))
    # simulate data within segment
    for (j in 1:nq) {
        # simulate normal distribution
        res <- rnorm(group_size[i], mean = mus2[i,
            j], sd = sd2)
```

```
        # set upper and lower limit
        res[res > 5] <- 5
        res[res < 1] <- 1
        # convert continuous values to discrete integers
        seg[, j] <- floor(res)
    }
    sim.dat2 <- rbind(sim.dat2, seg)
}


names(sim.dat2) <- paste("Q", 1:10, sep = "")
sim.dat <- cbind(sim.dat, sim.dat2)
sim.dat$segment <- factor(rep(group_name, times = group_size))
```

B.2 Swine Disease Breakout Data

```
# sim1_da1.csv the 1st simulated data similar
# sim1_da2 and sim1_da3 sim1.csv simulated data,
# the first simulation dummy.sim1.csv dummy
# variables for the first simulated data with all
# the baseline code for simulation

nf <- 800
for (j in 1:20) {
    set.seed(19870 + j)
    x <- c("A", "B", "C")
    sim.da1 <- NULL
    for (i in 1:nf) {
        # sample(x, 120, replace=TRUE)->sam
        sim.da1 <- rbind(sim.da1, sample(x, 120, replace = TRUE))
    }

    sim.da1 <- data.frame(sim.da1)
```

```r
col <- paste("Q", 1:120, sep = "")
row <- paste("Farm", 1:nf, sep = "")
colnames(sim.da1) <- col
rownames(sim.da1) <- row


# use class.ind() function in nnet package to encode
# dummy variables
library(nnet)
dummy.sim1 <- NULL
for (k in 1:ncol(sim.da1)) {
    tmp = class.ind(sim.da1[, k])
    colnames(tmp) = paste(col[k], colnames(tmp))
    dummy.sim1 = cbind(dummy.sim1, tmp)
}
dummy.sim1 <- data.frame(dummy.sim1)


# set 'C' as the baseline delete baseline dummy variable

base.idx <- 3 * c(1:120)
dummy1 <- dummy.sim1[, -base.idx]


# simulate independent variable for different values of
# r simulate based on one value of r each time r=0.1,
# get the link function

s1 <- c(rep(c(1/10, 0, -1/10), 40),
        rep(c(1/10, 0, 0), 40),
        rep(c(0, 0, 0), 40))
link1 <- as.matrix(dummy.sim1) %*% s1 - 40/3/10


# Other settings  ---------------------------
# r = 0.25
# s1 <- c(rep(c(1/4, 0, -1/4), 40),
#         rep(c(1/4, 0, 0), 40),
#         rep(c(0, 0, 0), 40))
# link1 <- as.matrix(dummy.sim1) %*% s1 - 40/3/4
```

```
# r = 0.5
# s1 <- c(rep(c(1/2, 0, -1/2), 40),
#         rep(c(1/2, 0, 0), 40),
#         rep(c(0, 0, 0), 40))
# link1 <- as.matrix(dummy.sim1) %*% s1 - 40/3/2


# r = 1
# s1 <- c(rep(c(1, 0, -1), 40),
#         rep(c(1, 0, 0), 40),
#         rep(c(0, 0, 0), 40))
# link1 <- as.matrix(dummy.sim1) %*% s1 - 40/3


# r = 2
# s1 <- c(rep(c(2, 0, -2), 40),
#         rep(c(2, 0, 0), 40),
#         rep(c(0, 0, 0), 40))
#
# link1 <- as.matrix(dummy.sim1) %*% s1 - 40/3/0.5


# calculate the outbreak probability
hp1 <- exp(link1)/(exp(link1) + 1)


# based on the probability hp1, simulate response
# variable: res
res <- rep(9, nf)
for (i in 1:nf) {
    res[i] <- sample(c(1, 0), 1, prob = c(hp1[i], 1 -
        hp1[i]))
}


# da1 with response variable, without group indicator
# da2 without response variable, with group indicator
# da3 without response variable, without group indicator

dummy1$y <- res
```

```r
da1 <- dummy1
y <- da1$y
ind <- NULL
for (i in 1:120) {
    ind <- c(ind, rep(i, 2))
}


da2 <- rbind(da1[, 1:240], ind)
da3 <- da1[, 1:240]

# save simulated data
write.csv(da1, paste("sim", j, "_da", 1, ".csv", sep = ""),
    row.names = F)
write.csv(da2, paste("sim", j, "_da", 2, ".csv", sep = ""),
    row.names = F)
write.csv(da3, paste("sim", j, "_da", 3, ".csv", sep = ""),
    row.names = F)
write.csv(sim.da1, paste("sim", j, ".csv", sep = ""),
    row.names = F)
write.csv(dummy.sim1, paste("dummy.sim", j, ".csv",
    sep = ""), row.names = F)
}
```

Bibliography

Albert, A. and Anderson, A. J. (1984). On the existence of the maximum likelihood estimates in logistic regression models. *Biometrika*, 71(1):1–10.

Efron, B. and Tibshirani, R. (1986). Bootstrap methods for standard errors, confidence intervals, and other measures of statistical accuracy. *Statistical Science*, 1(1):54–75.

Bauer, E. and Kohavi, R. (1999). An empirical comparison of voting classification algorithms: Bagging, boosting, and variants. *Machine Learning*, 36:105–142.

Ben-Dor, A., Bruhn, L., Friedman, N., Nachman, I., Schummer, M., and Yakhini, Z. (2000). Tissue classification with gene expression profiles. *Journal of Computational Biology*, 7(3):559–583.

Bergstra, J., Casagrande, N., Erhan, D., Eck, D., and Kegl, B. (2006). Aggregate features and adaboost for music classification. *Machine Learning*, 65:473–484.

Box, G. E. P. and Cox D. R. (1964). An analysis of transformations. *Journal of the Royal Statistical Society*, 26(2):211–252.

Breiman, L. (1998). Arcing classifiers. *The Annals of Statistics*, 26:123–140.

Breiman, L. (2001a). Random forests. *Machine Learning*, 45:5–32.

Breiman, L. (2001b). Statistical modeling: The two cultures. *Statistical Science*, 16(3):199231.

Breiman, L., Friedman, J. H., Olshen, R. A., and Stone, C. J. (1984). *Classification and Regression Trees*. CRC.

Cestnik, B. and Bratko, I. (1991). On estimating probabilities in tree pruning. *EWSL '91: Proceedings of the European Working Session on Machine Learning*, Pages 138–150.

Chollet, F. (2017). *Deep Learning with Python.* Manning.

Chollet, F. and Allaire, J. J. (2018). *Deep Learning with R.* Manning.

Chun, H. and Keleş, S. (2010). Sparse partial least squares regression for simultaneous dimension reduction and variable selection. *Journal of the Royal Statistical Society: Series B (Statistical Methodology)*, 72(1):3–25.

Chung, J., Gulcehre, C., Cho, K., and Bengio, Y. (2014). Empirical evaluation of gated recurrent neural networks on sequence modeling. *CoRR*, http://arxiv.org/abs/1412.3555.

McClish, D. (1989). Analyzing a portion of the roc curve. *Medical Decision Making*, 9:190–195.

DeLong, E R., DeLong, D. M. and Clarke-Pearson, D. L. (1988). Comparing the areas under two or more correlated receiver operating characteristics curves: a nonparametric approach. *Biometrics*, 44:837–845.

de Waal, T., Pannekoek, J., and Scholtus, S. (2011). *Handbook of Statistical Data Editing and Imputation.* John Wiley and Sons.

Dudoit S., Fridlyand J., and Speed T. (2002). Comparison of discrimination methods for the classification of tumors using gene expression data. *Journal of the American Statistical Association*, 97(457):77–87.

Efron, B. (1983). Estimating the error rate of a prediction rule: Improvement on cross-validation. *Journal of the American Statistical Association*, 78(382):316–331.

Efron, B. and Tibshirani, R. (1986). Bootstrap methods for standard errors, confidence intervals, and other measures of statistical accuracy. *Statistical Science*, 1(1):54–75.

Efron, B. and Tibshirani, R. (1997). Improvements on cross-validation: The 632+ bootstrap method. *Journal of the American Statistical Association*, 92(438):548–560.

Esposito, F., Malerba, D., and Semeraro, G. (1997). A comparative analysis of methods for pruning decision trees. *IEEE Transactions on Pattern Analysis and Machine Intelligence*, 19(5):476–491.

Freund, Y. and Schapire, R. (1997). A decision-theoretic generalization of online learning and an application to boosting. *Journal of Computer and System Sciences*, 55:119–139.

Friedman, J., Hastie, T., and Tibshirani, R. (2000). Additive logistic regression: A statistical view of boosting. *Annals of Statistics*, 38:337–374.

Gareth James, Daniela Witten, T. H., and Tibshirani, R. (2015). *An Introduction to Statistical Learning*. Springer, 6th edition.

Geladi, P. and Kowalski, B. (1986). Partial least squares regression: A tutorial. *Analytica Chimica Acta*, (185):1–17.

Gelman, A. and Hill, J. (2006). *Data Analysis Using Regression and Multilevel/Hierarchical Models*. Cambridge University Press.

Goodfellow, I., Bengio, Y., and Courville, A. (2016). *Deep Learning*. MIT Press.

Hall, P., Hyndman, R., and Fan, Y. (2004). Nonparametric confidence intervals for receiver operating characteristic curves. *Biometrika*, 91:743–750.

Hand, D. and Till, R. (2001). A simple generalisation of the area under the roc curve for multiple class classification problems. *Machine Learning*, 45(2):171–186.

Hastie, T., Tibshirani, R., and Friedman, J. (2008). *The Elements of Statistical Learning: Data Mining, Inference and Prediction*. Springer, 2nd edition.

Hochreiter, S. and Schmidhuber, J. (1997). Long short-term memory. *Neural Computation*, 9(8):1735–1780.

Hoerl, A. and Kennard, R. (1970). Ridge regression: Biased estimation for nonorthogonal problems. *Technometrics*, 12(1):55–67.

Hssina, B., Merbouha, A., Ezzikouri, H., and Erritali, M. (2014). A comparative study of decision tree ID3 and C4.5. *International Journal of Advanced Computer Science and Applications(IJACSA), Special Issue on Advances in Vehicular Ad Hoc Networking and Applications 2014*, 4(2). http://dx.doi.org/10.14569/SpecialIssue.2014.040203

Hyndman, R. and Athanasopoulos, G. (2013). *Forecasting: Principles and Practice*, volume Section 2/5. OTect: Melbourne, Australia.

Iglewicz, B. and Hoaglin, D. (1993). How to detect and handle outliers. *The ASQC Basic References in Quality Control: Statistical Techniques*, 16.

Cohen, J. (1960). A coefficient of agreement for nominal data. *Educational and Psychological Measurement*, 20:37–46.

Jolliffe, I. (2002). *Principal Component Analysis*. Springer, 2nd edition.

Kuhn, M. and Johnston, K. (2013). *Applied Predictive Modeling*. Springer.

Kwak, G. H. J. and Hui, P. (2019). Deephealth: Deep learning for health informatics reviews, challenges, and opportunities on medical imaging, electronic health records, genomics, sensing, and online communication health. *ACM Transactions on Computing for Healthcare.*

Breiman, L. (1966). Bagging predictors. *Machine Learning*, 24(2): 123–140.

Valiant, L. (1984). A theory of the learnable. *Communications of the ACM*, 27:1134–1142.

Meier, L. Geer, Sara van de, and Buhlmann, P. (2008). The group lasso for logistic regression. *Journal of the Royal Statistical Society Series B (Methodological)*, 70:53–71.

Lachiche, N. and Flach, P. (2003). Improving accuracy and cost of two–class and multi–class probabilistic classifiers using roc curves. *In "Proceed- ings of the Twentieth International Conference on Machine Learning*, 20:416–424.

Landis, J. R. and Koch, G. G. (1977). The measurement of observer agreement for categorical data. *Biometrics*, 33:159–174.

Li, J. and Fine, J. P. (2008). ROC analysis with multiple classes and multiple tests: Methodology and its application in microarray studies. *Biostatistics*, 9(3):566–576.

Line Clemmensen, Trevor Hastie, Daniela Witten, and Bjarne Ersbøll. (2011). Sparse discriminant analysis. *Technometrics*, 53(4):406–413.

Michael Kearns and Leslie Valiant. (1989). Cryptographic limitations on learning boolean formulae and finite automata. In *Proceedings of the Twenty-First Annual ACM Symposium on Theory of Computing*.

Maytal Saar-Tsechansky and Foster Provost. (2007). Handling missing values when applying classification models. *Journal of Machine Learning Research*, 8(57):1623–1657.

McElreath, R. (2020). *Statistical Rethinking: A Bayesian Course with Examples in R and STAN*. Chapman and Hall/CRC.

Mulaik, S. (2009). *Foundations of Factor Analysis*. Chapman Hall/CRC, 2nd edition.

Patel, N. and Upadhyay, S. (2012). Study of various decision tree pruning methods with their empirical comparison in weka. *International Journal of Computer Applications*, 60(12):20–25.

Pearl, J. and Mackenzie, D. (2019). *The Book of Why*. Penguin Books.

Provost, F., Fawcett, T., and Kohavi, R. (1998). The case against accuracy esti- mation for comparing induction algorithms. *Proceedings of the Fifteenth International Conference on Machine Learning*, pages 445–453.

Quinlan, J.R. (1987). Simplifying decision trees. *International Journal of Human-Computer Studies*, 27(3):221–234.

Tibshirani, R. (1996). Regression shrinkage and selection via the lasso. *Journal of the Royal Statistical Society Series B (Methodological)*, 58(1):267–288.

Serneels, S., Nolf, E. D., and Espen, P. V. (2006). Spatial sign preprocessing: A simple way to impart moderate robustness to multivariate estimators. *Journal of Chemical Information and Modeling*, 46(3):1402–1409.

Dieterich, T. (2000). An experimental comparison of three methods for constructing ensembles of decision trees: Bagging, boosting, and randomization. *Machine Learning*, 40:139–158.

Fawcett, T. (2006). An introduction to roc analysis. *Pattern Recognition Letters*, 27(8):861–874.

Ho, T. (1998). The random subspace method for constructing decision forests. *IEEE Transactions on Pattern Analysis and Machine Intelligence*, 13:340–354.

Varmuza, K., He, P., and Fang, K. (2003). Boosting applied to classification of mass spectral data. *Journal of Data Science*, 1:391–404.

Wasserstein, R. L. and Lazar, N. A. (2016). The ASA's Statement on p-Values: Context, Process, and Purpose. *The American Statistician*, 70(2):129–133.

Massy, W. (1965). Principal components regression in exploratory statistical research. *Journal of the American Statistical Association*, 60:234–246.

Wedderburn, R. W. M. (1976). On the existence and uniqueness of the maximum likelihood estimates for certain generalized linear models. *Biometrika*, 63:27–32.

Willett, P. (2004). Dissimilarity-based algorithms for selecting structurally diverse sets of compounds. *Journal of Computational Biology*, 6(3-4):447–457. doi:10.1089/106652799318382

Wold, H. (1973). Nonlinear iterative partial least squares (NIPALS) modelling: Some current developments. *Academic Press*, 383–407.

Wold, H. and Jöreskog, K. G. (1982). *Systems Under Indirect Observation: Causality, Structure, Prediction*. North Holland, Amsterdam.

Amit, Y. and Geman, D. (1997). Shape quantization and recognition with randomized trees. *Neural Computation*, 9:1545–1588.

Kim, Y., Kim, J., and Kim, Y. (2006). Blockwise sparse regression. *Statistica Sinica*, 16(2):375–390.

Yeo, G. and Burge, C. (2004). Maximum entropy modeling of short sequence motifs with applications to RNA splicing signals. *Journal of Computational Biology*, 11(2-3):377–394. doi: 10.1089/1066527041410418. PMID: 15285897.

Yoav Freund and Robert E. Schapire. (1999). Adaptive game playing using multiplicative weights. *Games and Economic Behavior*, 29:79–103.

Yuan, M. and Lin, Y. (2007). Model selection and estimation in regression with grouped variables. *Journal of the Royal Statistical Society Series B (Methodological)*, 68:49–67.

Zhang, C., Bengio, S., Hardt, M., Recht, B., and Vinyals, O. (2017). Understanding deep learning requires rethinking generalization. *arXiv:1611.03530*.

Zou, H. and Hastie, T. (2005). Regularization and variable selection via the elastic net. *Journal of the Royal Statistical Society, Series B*, 67(2):301–320.

Index

For Product Safety Concerns and Information please contact our EU
representative GPSR@taylorandfrancis.com
Taylor & Francis Verlag GmbH, Kaufingerstraße 24, 80331 München, Germany

www.ingramcontent.com/pod-product-compliance
Ingram Content Group UK Ltd.
Pitfield, Milton Keynes, MK11 3LW, UK
UKHW020932180425
457613UK00012B/326